HOW TO READ THE
SOLAR
SYSTEM

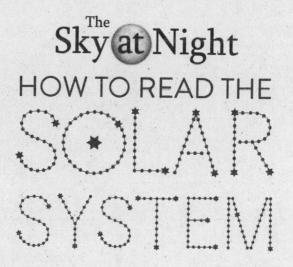

The Sky at Night
HOW TO READ THE
SOLAR SYSTEM

CHRIS NORTH AND PAUL ABEL

BOOKS

for Patrick

This book is published to accompany the BBC series entitled
The Sky at Night, first broadcast in 1957.

Series producer: Jane Fletcher
Executive producer: William Lyons

3 5 7 9 10 8 6 4 2

BBC Books, an imprint of Ebury Publishing
20 Vauxhall Bridge Road
London SW1V 2SA

BBC Books is part of the Penguin Random House group of companies
whose addresses can be found at global.penguinrandomhouse.com

Penguin
Random House
UK

First published by BBC Books in 2013
This paperback edition published in 2018

www.penguin.co.uk

A CIP catalogue record for this book is available from the British Library.

ISBN: 978 1 84990 629 6

Penguin Random House is committed to a sustainable future for
our business, our readers and our planet. This book is made
from Forest Stewardship Council® certified paper

MIX
Paper from
responsible sources
FSC® C018179

Printed and bound in Great Britain by Clays Ltd, Elcograf S.p.A.

Contents

Acknowledgements 9

Foreword 11

* 1 * What is the Solar System? 15

* 2 * Ancient Stargazers 23

* 3 * Celestial Mechanics 37

* 4 * The Sun 59

* 5 * Mercury 81

* 6 * Venus 95

* 7 * Earth and Moon 119

* 8 * Mars 151

* 9 * Jupiter 177

* 10 * Saturn 201

* 11 * Uranus and Neptune 233

* 12 * Asteroids and Dwarf Planets 253

* 13 * Comets 275

* 14 * One Among Many 295

Glossary of Useful Terms 305

Index 311

Picture Credits 320

Acknowledgements

It goes without saying that any book associated to *The Sky at Night* owes a huge debt of gratitude to Patrick Moore, who presented the show from its first airing in April 1957 until his death, at the age of 89, in December 2012. His personality, charisma and amazing ability to explain the most complex ideas in quick and easy to understand way has resulted in an amazing legacy of more than 55 years, and over 700 programmes. Patrick frequently insisted that *The Sky at Night* was always about more than just him, and while he is sorely missed, the series has continued.

Without the continuation of *The Sky at Night* this book would simply not have happened. A common question has been who would fill Patrick's shoes, and the honest answer is that no single person can and the presenting team numbers six: Chris Lintott, Lucie Green, Pete Lawrence, Jon Culshaw and the two of us. Added to that, of course, are the guests we feature every month, from professional astronomers to amateur astronomical societies; the contributions of their time, expertise and back catalogue of amazing images help make the programme a success. Coordinating the whole show are Jane Fletcher, our series producer, and Bill Lyons, the executive producer. We are grateful to both Jane and Bill, without whom the series would have almost certainly ended with Patrick, and who have both leant their support to this book. Jane and Bill are supported by a small but efficient production team based at BBC Bristol, in particular Alison Suker, Stella Stylianos and Keaton Stone. In the edit suite, Glenn Lewis, Dilesh Korya and Steve Williams have made the programme look and sound smooth and polished every month. The fact that this is possible in the space of a few days is testament to the skill and hard work of the film crew, particularly Andy Davis, Rob Lacey, Rob Hawthorne, Martin Huntley, Mark Payne-Gill..

We would also like to add a word of thanks to Richard Baum who was of great assistance for some of the historical sources in this book. And finally, it is important to say that such efforts wouldn't be possible without the support of our friends, family, and colleagues, who have understood when we've had to spend significant amounts of time researching and writing parts of this book. We are immensely grateful to Gabi, Clara and Matthew.

Chris North
Paul Abel
Autumn 2013

Foreword

How To Read the Solar System is a book that would have warmed Sir Patrick Moore's heart – for many reasons. It has a simple aim – to provide, for those just beginning in astronomy, a guide to the awesome motley collection of objects that make up the family of our own star – the yellow ball of incandescence which we call the Sun.

To call this assortment of rocks a 'family' is not quite as far-fetched as it may seem, for although they come in an array of vastly differing shapes and sizes, these lumps of matter all share a common history – the history of the solar system, and many of their births are related. The more we discover about these planets, moons, comets and meteors, the more surprises they reveal. Yet, on a cosmic scale, we are looking here at the tiniest portion of the known Universe – an almost infinitesimally small fraction of the vast expanse of space that we see looking up on a clear night, if we are lucky enough to be able to escape the all-pervasive light pollution of the modern world.

What we see in the heavens, even with the naked eye, is a myriad of stars – each one a Sun in its own right, and recent research indicates that most of them probably have families of planets of their own. This appears to be the norm, rather than the exception, so there are probably even more planets than there are stars out there.

With binoculars or a small telescope, more wonders are revealed. We are able to glimpse, from within, the structure of our own Milky Way galaxy, a kind of greater family into which

Sir Patrick Alfred Caldwell Moore (1923-2012).

all the stars in our skies fit. We can look at the brightest parts of the Milky Way and see the massive dense centre of our galaxy, and look in the opposite direction out into the outer reaches of its spiral arms. And looking further out from our galaxy we are even able to see other galaxies of similar size, each containing billions of stars of their own. The most powerful telescopes now available have peered into unimaginably distant space, and, because of the finite speed of their light coming towards us, seen billions of years back in time. What they have shown us is billions upon billions of galaxies stretching out to the limits of the observable Universe.

This is a very big picture, and the study of it, on this enormous scale, is called cosmology. Yet, this book has no truck with cosmology. It's impossible to see any of this vastness at close quarters, but many of us, including Sir Patrick Moore, have found great thrills in examining what is on our own doorstep, relatively speaking. Looking up at the skies of our own Solar System, we soon become aware of those 'wanderers' which move their position on the starry vault every night – the planets which have been known to our ancestors since before recorded history: Mercury, Venus, Mars, Jupiter, Saturn. And we may be lucky enough to see a visiting comet or a minute piece of rock flash its burning path through our atmosphere as it burns up – the 'shooting stars' which also fascinated our forbears.

It is this close-up drama that is the subject of this book. It is a modest aim, to survey only material within a couple of light years of our Sun, but in this study, we will see in miniature the whole of the Universe, since it seems that no matter where we go in the Universe things look very much the same. So a careful study of our immediate surroundings can tell us so much about the wider cosmos. Indeed, Sir Patrick Moore, though well versed in every aspect of current knowledge in astronomy and astrophysics, devoted a large fraction of his entire life to detailed observations of the planets, and in particular, our own Moon, a mere quarter of a million miles from Earth. His meticulous lunar mapping was so extraordinarily accurate that is was used by the first US astronauts, to plan their conquest of the Moon.

Sir Patrick was not the only astronomer to become absorbed in the uncovering the secrets of our Solar System, but our knowledge of this subject has exploded in the last 20 years, as a succession of unmanned probes have been launched on missions to rendezvous with the planets, their moons, and even with cometary nuclei and asteroids. The photographs and data these probes have sent back to Earth have brought us shockingly clear information about all these objects, and it is fair to say that the truth about the composition of all these objects was a surprise in every case.

Does that 'uniformity' we see in the Universe include life? Are those other planets inhabited too, with the wonderful array of creatures which Man has done such a good job of bringing to the brink of extinction? We may never know. But looking at those rocks and stones, and the very dust that surrounds us on our tiny blue planet is a very good place to start. Certainly for this Sir Patrick, to whom this admirable book is dedicated, and who is the reason so many of us have discovered the thrills of astronomy, would be smiling.

Dr Brian May

1

What is the Solar System?

We should probably start this book by stating what is meant by the Solar System, and what better place to look than the dictionary. The word 'solar' comes from the Latin *solaris*, which means 'of the Sun', and so we define the Solar System to be the collection of objects that 'belong to' the Sun – that is to say, everything that orbits it.

The Sun is the powerhouse of the whole Solar System, and presents a perfectly good place to start our introduction to the Solar System. Although it may be incredibly important to our brief lives here on Earth – and was portrayed by our ancestors as an all-seeing deity – the Sun is in fact more or less an ordinary star. But in astronomy, the word 'ordinary' is often overused. The Sun is over a million kilometres across, and weighs in at a whopping two thousand million million million million tonnes. Its surface is pretty hot by human standards, at around 5,500°C, but its centre is at a temperature of tens of millions of degrees – so hot and dense that nuclear fusion is converting four million tonnes of matter into pure energy *every single second*.

This incredible star, while not significantly different from the others in our Galaxy, is host to a magnificent array of planets,

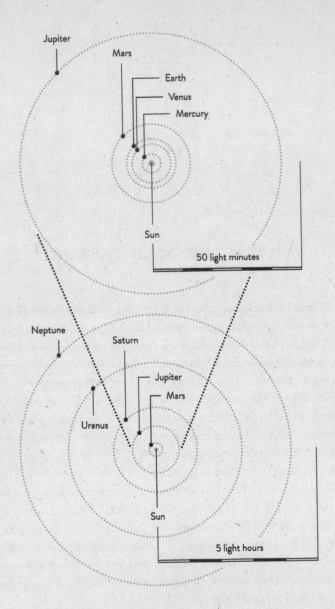

The orbits of the planets in our Solar System,
with distance scales shown in light travel time.

moons, asteroids and comets – including one very special planet that we call home. There are, of course, the eight major planets, which in order of increasing distance from the Sun are Mercury, Venus, Earth, Mars, Jupiter, Saturn, Uranus and Neptune. We often groups these into two main sections, the inner planets (Mercury through to Mars) and outer planets (Jupiter to Neptune), and there are useful distinctions to be made at this level.

THE INNER PLANETS

The inner planets are, on the face of it, not too unlike the Earth – hence their other common name of 'terrestrial planets', with the word terrestrial originating from the Latin *terrestris*, or 'earthly'. They are largely made of rocky material, and are thought to have formed from the collision of lots of smaller pieces of space rocks around four and a half billion years ago. They are all – aside from the odd mountain range and deep canyon – almost completely spherical, with diameters ranging from 5,000 km (3,000 miles) for the tiny Mercury up to 13,000 km (6,000 miles) for the Earth.

Beyond that, however, these four worlds are all very different, and each has its own unique characteristics. Let's start with something we're all familiar with – Earth. While we inhabitants of the planet are aware of temperature variations that seem huge to us, these variations are tiny in the context of other planets and stars. In those terms, at an average of 150 million km (93 million miles) from the Sun, the Earth's surface temperature never moves too far from 0°C, allowing it to host rivers, lakes and oceans of liquid water. Its atmosphere is primarily a mixture of nitrogen and oxygen – a combination that supports a diverse ecosphere. Beneath the surface it has a molten mantle and a dense iron core, with geological features such as volcanoes dotted over its surface.

The least unlike the Earth of the other terrestrial planets is probably Mars, although it is only half the diameter. It is also further from the Sun than the Earth, at a distance of 225 million km (140 million miles) and its maximum surface temperature is just 35°C. Its atmosphere is much thinner, and composed primarily of carbon dioxide, and so is far less conducive to life. With such a thin atmosphere, and temperatures that can drop as low as -110°C at the poles, there is no standing liquid water on Mars – though there is lots of water ice both on and beneath the surface, and evidence that small amounts can flow on the surface for brief periods. It is not inconceivable that people could survive on Mars, although a spacesuit would certainly be required. There is also evidence of ancient volcanoes – Mars is host to the largest volcano in the Solar System – though they are long extinct. Its small size means that the interior of Mars cooled long ago and stopped all geological activity.

Our nearer neighbour, Venus, is about the same size as the Earth – but that is where the similarities cease. Like the porridge in *Goldilocks*, while the Earth is just the right temperature to support life and Mars is too cold, Venus is far, far too hot. Although it is closer to the Sun, at 108 million km (68 million miles), the main reason for its high temperature is its thick atmosphere. At some point in its past Venus experienced a runaway greenhouse effect, and its thick, sulphurous atmosphere bakes its surface to a staggering 450°C. Not only would anyone exploring Venus be boiled alive, and their craft gradually melted by the rain of sulphuric acid, they would also be crushed, as the thick atmosphere creates a surface pressure almost 100 times higher than Earth's.

Finally, we come to Mercury, the real oddball of the inner planets. It is the smallest of the four, the closest to the Sun – at an average distance of 58 million km (35 million miles) – and also the only one without an atmosphere. This means there is nothing to protect the surface from the Sun's glare, or to retain heat overnight. Mercury's surface drops to a chilly -180°C in the middle of the night at the poles, but reaches a searing 400°C at midday on the equator.

THE OUTER PLANETS

The inner planets may be weird and wonderful, but in comparison the outer planets are truly strange. The two largest, Jupiter and Saturn, are true giants, with Jupiter having more than ten times the diameter of the Earth. But neither is a solid ball of rock – both are composed almost entirely of gas, and may not even have a solid surface at all. They have storms in their atmospheres that could swallow the Earth, most notably Jupiter's Great Red Spot. Saturn is also accompanied by an impressive system of rings, made of billions of small icy particles, giving it perhaps the most iconic appearance of all the planets. As interesting as the planets themselves are their families of satellites, which number in the dozens. Some of these icy moons are planet-sized – Jupiter's largest moon, Ganymede, is larger than Mercury – and many of them have their own fascinating stories.

Beyond Jupiter and Saturn we have the ice giants, Uranus and Neptune. Although smaller than Jupiter and Saturn, they are still around four times the diameter of Earth. Beneath their thick atmospheres they are thought to have solid, icy cores. At distances of more than two billion km from the Sun, these planets are the ones we know least about, though what we do know tells us that they have their own intriguing characteristics and histories.

There is far more to say about these eight major planets and their moons, but there are also a whole host of smaller bodies is orbit around the Sun. First there are the minor planets: lumps of space rubble ranging in size from the size of pebbles up to dwarf planets hundreds of kilometres across. Most of these reside in the asteroid belt, between the orbits of Mars and Jupiter, and the Kuiper Belt, out beyond Neptune. Then there are comets, dirty balls of ice which lose matter as they come close to the Sun, forming beautiful tails of material. Though beautiful, the tails of comets are incredibly tenuous and only visible because the material they're made of reflects light so well. Throughout this book we will come to each of these objects in turn, though we will not be able to discuss each asteroid and comet individually as they number in the hundreds of thousands.

HOW DO WE KNOW ALL THIS?

There is much more to be said about all of these objects, but the next question that is commonly asked about our understanding of the Solar System is: how do we know all this? We haven't sent people to all these places, as the distances are simply too vast. The furthest a human being has been from the Earth is just over 400,000 km (250,000 miles), a feat achieved by the Apollo 13 astronauts on their ill-fated and near-disastrous trip to the Moon in 1970. Today, astronauts venture only as far as low Earth orbit, just a few hundred miles above the surface. Compare those distances with a minimum of 75 *million* kilometres to Mars, or more than a *billion* kilometres to Saturn, and the scale of the problem of human exploration becomes apparent.

Astronauts on the International Space Station regularly spend months in space, but the longest trips away from the relative safety of Earth orbit have lasted for less than two weeks. By comparison, a journey to Mars typically takes around 7 months – and that doesn't include any time on the surface or even a return trip. The amount of food, water, air and so on required for such a long trip, which in practice would have to take two years, is immense, and we have not yet managed to prove that we can protect explorers from the hazards of interplanetary travel. Deeper exploration of the Solar System would take longer still, with one-way trips to the outer planets taking several years. Even the fastest-launched spacecraft, New Horizons, will have been travelling for just shy of a decade by the time it passes Pluto.

All these, so far insurmountable, obstacles mean that we can't send people to the planets, though we can do the next best thing – send robots. Even then, the missions still take years to complete and cost millions of pounds. The results of these missions are fascinating, and well worth the time, effort and money, but it is important to remember that they are building on the successes of observations made right here on planet Earth.

Although modern discoveries about the planets, their moons, and the myriad of other objects in the Solar System tend to be made by expensive telescopes and space probes, many of the key observations about the Solar System can be achieved with much more modest equipment. After all, it must be remembered that much of the celestial mechanics was worked out by astronomers such as Nicolaus Copernicus and Tycho Brahe. They made observations using just their own eyes – combined with a large degree of patience and a very methodical approach to observing. Even after the development of the telescope, the likes of Galileo Galilei, Johannes Kepler and Thomas Harriot used telescopes much less powerful than the smallest astronomical telescopes on sale today.

The discoveries of these great astronomers were largely due to their diligence and dedication to their science, the attention and rigour they paid to the detail of making observations, and in some cases the courage to stand up to the doctrine of their time. But astronomy is far more than catalogues of numbers and observations. The great early astronomers were making observations because they wanted to understand the mechanics of the Universe, and they were driven by a strong sense of curiosity. Astronomy is still doing that, with observations from modern experiments and telescopes continuously updating our view of our place in the Universe. On top of that, our own Solar System can provide some of the most stunning and breathtaking views one is ever likely to see – all it needs is an appreciation of what one is looking at.

2

Ancient Stargazers

A dark night sky is a spectacular sight. Among the best places to view the night sky are the deserts of North America. On a moonless night, the dark velvety backdrop of sky is quite literally powdered with thousands of stars. Indeed it can be hard to make out the constellations when there are a bewildering number of suns present. In summertime, one of the spiral arms of the Milky Way galaxy within which our Sun resides stretches high overhead and looks bright enough to cast shadows. The scene has an almost three-dimensional quality to it, and is as close as a human being can come to touching the face of infinity.

Standing there in the desert, it is not hard to imagine our ancient ancestors hundreds of thousands of years ago, looking up from a similar spot, mesmerised by the silent spectacle above. Only the faint glow of the distant cities on the horizon hints at the technological presence of a species which only recently, by cosmological standards, had cowered in caves under these same stars.

It seems humans are endowed with a compulsion to record not just their own activities, but things that affected their lives. It is no surprise that it wasn't too long ago that mankind started to record the stars. The walls of the well-known Lascaux caves in south-west France are covered in many hundreds of beautiful images laid down by human beings some 17,300 years ago. Among the many images are some of the earliest drawings of the stars. In one particular painting, the eyes of a bull, a bird and a bird-man are believed to represent the three bright stars of the summer triangle, Vega, Deneb and Altair. Another painting shows a figure who seems to be associated with the Pleiades – indisputable proof of how the glory of the heavens captured our imagination at a very early stage in our evolution. It would start humanity on a path of exploration that would begin with mythology and storytelling, but would eventually lead to an understanding of the clockwork precision of the celestial machinery of nature.

A THEATRE IN THE SKY

If you go out on a dark night and look up, even if you live in a town or city, you will still see a number of stars. To our eyes, they form patterns known as constellations. There are 88 constellations in the night sky spread between the northern and southern hemispheres. From our point of view, it takes the stars a long time to move through space, and so as a result a constellation of stars may look the same for many thousands of years.

In the northern hemisphere, these constellations, whose names we use today, were established in Classical times and many of them are associated with various myths of the Ancient Greek period. What is interesting, though, is that not every culture saw the same thing, or used the same stars of a constellation consistently.

Perhaps one of the most recognisable constellations in the northern hemisphere is the Plough, or Great Bear. The main part of this constellation looks like a saucepan; indeed the French called it La Casserole. In Greek mythology, the god Zeus fell in love with a nymph called Callisto. Naturally, his wife was rather enraged at this and, out of jealousy, turned Callisto into a bear. Callisto's son, Arcas, was out hunting one day and was about to kill the bear that was really his mother; in order to avoid tragedy, Zeus put the bear into the night sky whereby it became the constellation of Ursa Major.

The people of Burma had a different name for this constellation. They called it *Pucwan Tārā* (pronounced 'bazun taja'), and to them the stars represented a prawn. From medieval England, people saw yet another pattern. To many of them it represented 'Charles Wain' (a wagon), and later a horse-drawn plough.

Perhaps the most imaginative visualisation of this constellation comes from the Chinese, who had an entirely separate system of constellations. For them, this group of stars represented a celestial bureaucrat who made his eternal rounds of the night sky and was followed by his two loyal petitioners.

Another striking constellation of the northern hemisphere is the constellation of Orion. In Greek mythology he represents a mighty hunter, but to the Ancient Egyptians the constellation represented the god Osiris, who was killed by his evil brother Set.

The Yognul people of Australia thought the constellation represented a canoe. But as we move into the southern hemisphere, a marked change in the nature of the constellations becomes apparent. The southern constellations were named by explorers of the 16th and 17th centuries. Instead of great heroes and gods, we have the keels of ships, sextants, telescopes and the like, things that mattered most to the people of this era.

What constellations would we put in the night sky today? Indeed, when human beings go out and colonise the stars, what constellations will they put up in their night skies? It would be wonderful to know what constellations the Sun will take part in thousands of years from now.

The Great Bear, or Ursa Major.

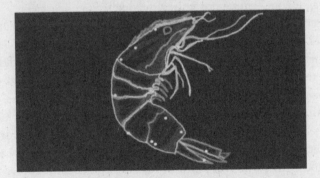

The Burmese identify Ursa Major
with a prawn or shrimp.

The Celestial Bureaucrat of Chinese mythology,
followed by his two faithful attendants.

THE DAWN OF SCIENCE

At first glance, the Universe looks chaotic. The rising and setting of the Sun and Moon, the motion of the planets, all seem somewhat random. However, as we begin to record the times and positions of these objects, we slowly become aware of the underlying clockwork of nature. Having a large number of observations allows us to see the patterns and cycles of nature and, if we can understand them, and find a way to formulate them, we can start to predict them.

Eventually, humans began to do more than just populate the skies with myths and legends – they began to record what they were seeing. The ancients were familiar with a number of events which regularly occurred in the night sky, and noted five bright 'stars' which, unlike the other stars in the sky, wandered around the constellations of the zodiac. These were called *planets*. There were also solar and lunar eclipses, which were both spectacular and for some cultures foretold of dire things to come.

Although we would regard the act of recording observations and modelling our results as scientific process, a number of cultures actually did this for superstitious reasons. Many of the great stone circles (like Stonehenge) have an astronomical connection and record the positions of the rising and setting of astronomical bodies at certain times of the year, but they also have religious aspects. Astronomy was presumably being used for the purpose of divining the future. It took a surprisingly long time for science to break free from the chains of superstition.

Perhaps the earliest recorded observations come from Bronze Age China. Dating from between 1400 and 1200 BC, the Shang dynasty oracle has inscriptions which seem to include references to stars and eclipses. These observations were made as it was believed they provided omens which foretold the state of the kingdom. Similarly, although the Ancient Egyptians built monuments to mark

A Babylonian stone tablet.

the rising and setting of certain stars, this also was for religious reasons connected with the rituals of the Pharaohs.

The Babylonians of ancient Mesopotamia do seem to have recorded their observations out of interest and we find a good deal more textual evidence for systematic observation of the night sky all wonderfully laid down in stone tablets. A number of texts dating from 650–50 BC contain observations of the time of sunrise, the times of the waxing and waning crescent moon, and details of lunar and solar eclipses.

By the 6th century BC, the Neo-Babylonians had begun to notice the patterns and repetitions in their observations. From their simple naked-eye, but faithful, recordings, they were able to calculate the time intervals between moonrise and moonset, the times of sunrise and sunset and the Moon's daily motion through the stars for many months ahead. They were also able to determine the length of daylight which could be expected, and the position of the Sun and Moon at the time of New Moon – an amazing achievement when you realise these people had no telescopes and no clocks. Perhaps for the first time in human history, the foundations of mathematical astronomy had been established.

The Classical Greeks made many of the more important contributions to science in antiquity. They were able to determine a great deal about the Solar System. The most surprising thing about their achievements is that all of their discoveries were made using a combination of simple geometry and naked-eye observations. We often think of the people of antiquity as somewhat primitive creatures. In the 21st century, technology is used as a defining point of civilisation, and it is through this lens of achievement that we often look back to judge the past. In fact, the people of antiquity were every bit as smart as 21st-century humans – biologically speaking, our brains would have been identical, and what they lacked in technology, they certainly made up for with scientific ingenuity.

HIPPARCHUS AND THE PRECESSION OF THE EQUINOXES

Perhaps the greatest Greek observational astronomer of that time was Hipparchus (190–120 BC). Hipparchus was both an astronomer and a superb mathematician. He is recognised as a founder of trigonometry (the study of right-angled triangles and the relationships between their internal angles and the lengths of their sides); he applied his geometry skills to his naked-eye astronomical observations, and in doing so was able to make a number of revolutionary discoveries about the Solar System. In his use of mathematics to explain his astronomical observations, Hipparchus is almost indistinguishable from today's modern astronomers. The problems modern astronomers work on may be more advanced, and the maths certainly is, but the fundamental idea that the workings of the Solar System can be understood by combining observation with mathematics is essentially the same. This concept was intuitively understood by Hipparchus some 2,000 years ago.

Hipparchus was born in the ancient city of Nicaea, which can be found today in the Turkish city of Iznik. In spite of his many great achievements, not much is known about him or how he supported his astronomical endeavours financially, and very little of his published work has survived. However, even though much of his work was lost, it had made enough of an impact to be recorded by other astronomers and philosophers of the period.

Hipparchus also studied the motion of the Sun and Moon. As seen from the Earth, the Moon orbits the Earth roughly once every 27 days. During this time, the Moon passes by many background stars – sometimes it even passes in front of them and this is called an occultation. This leads to an important phenomenon known as a *parallax*. There is a wonderful exercise that Patrick Moore used to demonstrate the phenomenon (you might want to do this at home rather than in public!). First of all, close one eye. Now, hold a finger up towards a distant object like the wall of the room. If you now look through the other eye, you will see that your finger appears to move with respect to the wall, and if you alternate between looking through your left and right eye, you can see your finger appears to move back and forth. This is a parallax, and the amount of motion is called a parallax angle and is measured in degrees.

Now if you were to replace your finger with the Moon and the wall with the background stars, and if instead you look at the Moon from two different places on the Earth's surface (rather than just changing eyes), then the Moon also undergoes a parallax with respect to the background stars, and the lunar parallax angle can be measured. How does this help? Well, by itself the observation doesn't mean much. If, however, we apply some basic trigonometry then we can get an estimate of the distance from the Moon to the Earth.

Perhaps Hipparchus's most outstanding achievement was his discovery of the precession of the equinoxes. We have two equinoxes, one in spring and one in autumn. At these times, if you were to stand on the Earth's equator, the Sun would be directly overhead at the zenith. At this time in spring, the Sun is in the constellation of Pisces; at the corresponding autumn equinox, the Sun is in Virgo. Now, the Earth has an axial tilt, and over the course of many thousands of years the direction the planet's axis points in changes. This is called precession. Currently, the north pole star is Polaris in Ursa Minor, but in 3000 BC it was the star Thuban in Draco. In AD 10,000 it will be the bright star Deneb in Cygnus, and eventually will return once more to Polaris. This slow precession not only changes the pole star, but also means the position of the equinoxes changes over time.

Hipparchus made measurements of the longitudes of a number of bright stars including Spica, the bright bluish star in Virgo. To his surprise, when he compared these observations to those made by astronomers over a hundred years previously, he found that Spica had shifted about two degrees with respect to the position of the autumn equinox. He also observed the length of a tropical year – the time it takes the Sun to return to the same point was different in length to a *sidereal* year (the time it takes the Sun to return to the same point in the sky). These observations led him to conclude that the position of the equinoxes changed slowly over time. He estimated one degree per century, astonishing accuracy for a man armed only with eyes and trigonometry – the actual figure is 1 degree 38 minutes.

In later years, Hipparchus made a catalogue of many hundreds of bright stars. This was not simply a list of stars; it also contained an estimate of their brightness. Astronomers use the term *magnitude* to mean how bright an object is in the sky. Hipparchus devised a system where bright stars were given magnitude 1, slightly fainter ones were assigned magnitude 2 and so on, all the way down to stars which were just visible to the naked eye, which were given magnitude 6. Remarkably, Hipparchus's magnitude system still remains in use today, although modern astronomers have updated it. For three centuries, Hipparchus's work would remain the dominant work in astronomy until the arrival of Claudius Ptolemy more than two centuries later.

Moving forward a little in time, there is another Greek philosopher who deserves a mention, Eratosthenes of Cyrene. Like so many people of that era, he was something of a renaissance man. He made many contributions to mathematics, geography and poetry. He was known affectionately as Beta, the second letter of the Greek alphabet, as it was claimed that *he* was the second best at everything.

A NEW SCIENCE

The great Library of Alexandria was constructed some time in the 3rd century BC, and was *the* seat of learning – an enormous centre for academic study in the ancient civilised world. It was a remarkable achievement and had collections of scrolls (called books) from many diverse places. It was the showpiece of Egypt, and Eratosthenes became its third librarian.

Eratosthenes had noticed something interesting. At noon, under the scorching sun on the day of the summer solstice, the Sun appeared to be directly overhead in the ancient city of Syene (now modern-day Aswan). If you were to stand a stick vertically in the ground, it would cast no shadow. However, a few miles away in the city of Alexandria, at precisely the same time, a vertical stick stuck in the ground did cast a small shadow. The only answer that would fit these observations was that the surface of the Earth was curved.

Eratosthenes measured the angle of the Sun's elevation at Alexandria at noon on the solstice and found it to be one fiftieth of a circle. He knew the distance to the city of Syene was around 5,000 *stadia*, which meant that one degree was equivalent to about 700 *stadia*. Although there were different values of the length of the *stadia*, if we use the Egyptian value then Eratosthenes' simple calculation gives the circumference of the Earth as 39,690 km. Astonishingly this value is only some 2 per cent off the true value. Like the Babylonians before him, Eratosthenes was able to make a significant discovery using little more than some simple solar observations, sticks and numbers.

Eratosthenes and Hipparchus were joined by many other Greek philosophers and scientists who actively tried to understand the physical world around them without a religious framework. The Mediterranean became the birthplace of the scientific process, of using rational arguments and reason to explain what was being observed. What the Greeks wanted to know was how the physical world fitted together in an overall pattern. What they wanted was a new science: the science of cosmology.

As you might expect, there were many early attempts to explain the Universe and our place within it. One early cosmological theory came from the Greek philosopher Anaximander (611–546 BC). Anaximander suggested that the Earth was flat and circular. Above the surface was the air, and clouds. He believed that all of the celestial bodies, the Sun, Moon, stars and planets, were holes in vast moving rings and it was the light from a fiery region behind these rings which shone through the holes.

It was Pythagoras who laid down the real beginnings of cosmology, setting down the basics of a cosmological theory whose underlying elements would last until the Renaissance. Pythagoras (580–500 BC) was an Ionian philosopher and mathematician. He founded a philosophical brotherhood in a Greek colony located in southern Italy. The Pythagoreans (as they were known) were somewhat in awe of what they believed to be the deep symbolic significance of geometry in nature.

They applied geometry to obtain estimates of the distances to various celestial objects and developed the foundations of what would eventually become Ptolemaic cosmology. In the Pythagorean model, the Earth was placed at the centre of the Universe. They believed that the Sun, stars, Moon and planets were attached to celestial spheres orbiting the Earth.

Interestingly, there was a very early attempt to remove the Earth from the centre of the Universe. Aristarchus of Samos (310–230 BC) used very basic geometry to estimate the sizes and distances of the Sun and the Moon. In his cosmological model, the Sun, not the Earth, was at the centre of the Solar System. This brave attempt at the right answer was not widely accepted; it simply made too much sense both for religious reasons and observational ones to have the Earth at the centre of the Universe.

Many of the objects in the sky – the Sun, the Moon and the stars – seemed to fit well into Pythagoras's model, but the planets stubbornly refused to behave appropriately. Eudoxus of Cnidus was the first to try to explain this. He wanted the cosmological model at the time to explain why the planets moved in the way they did, and he wanted to duplicate planetary motion using geometry.

The reason why Pythagoras's model could not accurately repeat planetary motion was that there were two major flaws in his theory. We know that the Earth is not at the centre of the Solar System, and the planets do not have circular orbits; instead they are elliptical. However, the circle was seen as divine, and so there was no question of having anything less than that in the heavens above. In order to make an Earth-based Solar System model with circular orbits work, Eudoxus added more and more spheres to make his approximations of reality better. In total, he added 55 spheres to make the model agreeable with the observed motion of the planets. Even with this vast number of spheres, the approximation still wandered. However, Claudius Ptolemy, a 2nd-century AD astronomer from Alexandria, was about to make the final refinements.

ORDERING HEAVEN:
THE PTOLEMAIC UNIVERSE

Like all of his predecessors, Ptolemy wanted to keep the circular orbit. As before, he also kept the Earth at the centre of the Solar System, with the planets orbiting the Earth. His model, though, has a marked difference from previous ones. Instead of vast numbers of crystal spheres, Ptolemy's model used epicycles.

The diagram overleaf shows the principles behind Ptolemy's epicycles. At the centre we have the Earth. The planet moves on an epicycle centred on the point E, which itself is orbiting the Earth on a path called a *deferent*. This geocentric model of the Universe was reasonably good at explaining the Solar System.

Ptolemy put his cosmological model together with his other works and produced *The Almagest*, a mathematical and astronomical treatise. It was the greatest astronomical textbook of the day, and would remain so for over 1,200 years. The treatise is a remarkable work, and its thirteen books cover Ptolemy's cosmology model, an introduction to spherical trigonometry,

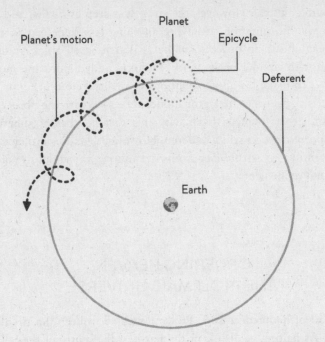

Planet's motion

Planet

Epicycle

Deferent

Earth

An epicycle used in Ptolemy's model of the Solar System.

lunar parallax, eclipses and observations and examinations of the orbits of the five bright planets.

The Almagest also contains a number of Hipparchus's discoveries, including the precession of the equinoxes and a list of some 1,022 stars and their estimated magnitudes. It wasn't only scientists and philosophers of the day who adopted this text and its ideas; many regions also followed suit. A geocentric cosmology with the planets moving on divine circular orbits seemed to be an obvious fingerprint of God, and to argue otherwise would become a dangerous path to heresy.

* 3 *

Celestial Mechanics

Since the 2nd century AD, the work of Claudius Ptolemy had been mankind's best effort to explain the Universe. This geocentric model, which had the Earth situated at the centre of the Solar System, with the Sun, Moon and planets orbiting it, had been universally adopted. The idea seemed to make sense from both an observational and religious point of view. As empires rose and fell, the Ptolemaic theory of the Solar System survived, unchallenged.

15th-century Europe was an interesting place. It was a place of revolution and war, but it was also home to free thinking and the foundations of modern science. It marks the period of the late medieval period with the promise of the early Renaissance to follow. Very shortly, the human race would pull itself out of the Dark Ages, out of a time when superstition and fear ruled the everyday lives of ordinary people and noblemen alike. For the Renaissance to take place, brave new ideas about society, the world, and its place within the Universe would be needed. By shining a light on the clockwork mechanisms of nature, science provided the human race with a means of escape from a world governed by demons and witches. It would in time provide an escape from the planet itself.

THE DAWN OF REASON

It was into this world, on 19 February 1473, that the Polish astronomer Nicolaus Copernicus was born in Torun, Poland. For the time, his family were quite wealthy, and Nicolaus was the youngest of four children. It seems he went to a number of schools, first St John's School in Torun, then later the Cathedral School at Włocławek. Afterwards, he entered the University of Krakow.

Copernicus studied a number of subjects at Krakow including art and astronomy. It was here that he learned the essential mathematical tools of arithmetic computation, geometry and algebra, skills that would be essential to his revolutionary theories later on. The young Copernicus was also a great reader. He read the works of Aristotle, and became familiar with the cosmology which had been in place for thousands of years, unchallenged and unchecked. Copernicus also read many of the other ancient Greek philosophers including Euclid, who had, thousands of years before, laid down the mathematical foundations of geometry.

It seems that Copernicus never finished his degree at Krakow. We know that he left there some time in 1495 and went to stay with his uncle, who had recently become the Prince-Bishop of Warmia. In January 1497, Copernicus attended Bologna University where he was to spend the next three years studying, perhaps not very enthusiastically, canon law. Copernicus was still interested in astronomy; the books of Aristotle and Euclid had prompted him to think about astronomical concepts for himself. While at Bologna, he made a number of observations of the Moon's path through the constellation of Taurus the Bull. He discovered that there were a number of discrepancies in Ptolemy's model which did not properly account for lunar motion. It was perhaps around this time that Copernicus began thinking about structuring the Solar System in a radically different way.

After a further spell at both Padua and Ferrara universities, where he studied medicine, he finally returned to Warmia, where he worked as an ecclesiastical administrator for his uncle. He lived at the Bishop's castle, and although his administrative duties kept him busy, Copernicus still found the time to work on his new model of the Solar System.

At some point before 1514, Copernicus set forth a rough draft of his ideas in a 40-page pamphlet called his *Commentariolus* (a little commentary). The pamphlet laid down the foundations for a new interpretation of the Solar System: – at its centre was the Sun, and all of the planets, including the Earth, moved around it in circular orbits. Copernicus also introduced another new concept: in the old Ptolemaic system, the external Universe revolved around the Earth once a day – in Copernicus's new model, the motion of the stars in the sky was due to the Earth turning on its axis once a day. This new model was the heliocentric model of the Solar System – *helios* is the Greek word for 'sun-centred'. What this draft lacked in mathematical framework it more than made up for with bold originality. The pamphlet was circulated amongst Copernicus's friends and colleagues; it was a brave step to challenge the well-established ideas of Ptolemy, and the pamphlet allowed Copernicus to gauge the reaction to his new ideas.

Copernicus was not only an excellent theorist, he was also an avid observer and, even though he had a heavy workload, he was still able to make many observations of the motions of the Sun, Mars and Saturn. He also began to construct a mathematical framework for his heliocentric model. This is essential: an astronomical theory must be tested to see if it is accurate, and the only way to test an idea is to examine closely the predictions it makes. Words are not enough; numbers, predicted positions and suggested planetary motions are the backbones of this type of astronomy.

Word began to spread of the new heliocentric system, and a number of people were keen to learn more about it. Copernicus, however, was a cautious man, and it seems that he had some doubts and anxieties about publishing his new ideas more fully. Perhaps he was anxious to have a proper mathematical framework, or maybe he was simply worried that his ideas would fall on stony ground; perhaps he feared ridicule. Whatever the reason, Copernicus chose not publish his theories for another 30 years. We may look back at this decision now with bewilderment, knowing that Copernicus's heliocentric model is essentially correct; indeed, it was about to ignite the new science of celestial mechanics. From Copernicus's point of view however things were very different. This was an age of limited communication, and a person's character and standing in the community was everything. It is no wonder he proceeded with caution.

Eventually, however, Copernicus did start to put all his ideas into a comprehensive text. This text, largely completed by 1532, bore the languid title *De revolutionibus orbium coelestium* ('On the Revolution of the Heavenly Orbs'). We have the Lutheran astronomer Georg Joachim Rheticus to thank for this work ever seeing the light of day. Rheticus was a cartographer and astronomer, and in May 1539 he visited Copernicus and became his student. Rheticus studied the new heliocentric model and realised its potential. He nagged and pushed the cautious Copernicus into publishing his work so that others could examine it properly.

The final form of 'On the Revolution of the Heavenly Orbs' contains seven new fundamental points. They were groundbreaking, revolutionary and clearly the product of a fine enquiring mind:

1. There is no centre of all the celestial spheres and circles.
2. The centre of the Earth is not the centre of the Universe. The Earth is only the centre of the Earth–Moon system.
3. All of the planets orbit the Sun in circles, with the Sun at the centre of the Universe.

4. The external Universe (i.e. the stars) is a vast distance from the Solar System.
5. The motion of the stars in the sky is entirely due to the Earth's rotation. The Earth revolves on an axis which passes through the north and south poles.
6. The motion of the Sun in the sky is entirely due to the rotation of the Earth and the motion of Earth in its orbit around the Sun.
7. The retrograde motion of the planets (i.e. when they appear to move backwards along their paths in the sky) is caused by the Earth's motion around the Sun.

Perhaps just as important as putting the Sun at the centre of the Solar System was that final point in Copernicus's text – a point which later would lead Johannes Kepler to the completion of his laws of planetary motion.

Ptolemy's ideas had been published in his *Almagest*, which had been the bible of astronomy since the 2nd century. Copernicus's work was its worthy replacement, and it contained the seeds of a new astronomy. It was published in a final form in 1643, by which time Copernicus was on his deathbed, but it is said that he was presented with a copy of his book shortly before he died.

Heliocentric cosmology soon made its way out into Europe, but it did not meet with universal support. There was a problem brewing for the supporters of heliocentrism; the teachings of Ptolemy fitted well with the Christian view of the Universe – it seemed so natural that God would put the Earth at the centre of the Universe, with all creation majestically dancing around us for eternity. Initially it seems that the Church was unaware of the dichotomy between the two systems. However that was about to change when the new heliocentrism became entangled with the life of a heretic and was used as a defence at the Inquisition. The ordeal would end with the heliocentric model being considered a heresy.

To begin with, the main objection to the ideas of Copernicus came from astronomers themselves. It seemed nonsensical to most of them that the Earth moved around the Sun; surely we would be able to feel this rapid motion if it were true, they reasoned, and, since the Church was unaware of the incompatibility of Copernicus's ideas with the established wisdom of the time, the idea of a heliocentric Solar System drifted around Europe as a mild curio. The main purpose of astronomy at that time was the prediction of the position of the planets and objects in the sky, not necessarily a complete physical understanding of why the Solar System worked the way it did.

SPREADING THE WORD:
THE LIFE OF A HERETIC

Giordana Bruno, who was born Fillippo Bruno in Campania in 1548, was not an astronomer, but he was a great advocate of the ideas of Copernicus and of free thinking. His support for the heliocentric Solar System was perhaps a double-edged sword.

By the age of 17, Bruno had entered the Dominican Order at the monastery of San Domenico in Naples, where he took the name Giordano. By the age of 24, he had become a fully qualified priest. One of Bruno's many skills was his incredible memory, and he developed a number of memory techniques over time, using mnemonics. He then developed a taste for free thinking that would ultimately lead to his downfall, and became interested in books the Church had listed as forbidden. Bruno had acquired the writings of Erasmus and seemed to embrace Arian Catholocism. His free thinking and subtle transgressions of monastery life and rules were not too severe, but owning a forbidden book was a serious thing. When Bruno discovered that the authorities were planning to investigate him, he fled Naples.

Between 1576 and 1583, Bruno became something of a nomadic wanderer. His meanderings took him to Noli, Turin, Venice and other places. After fleeing Naples it seems he abandoned his religious gowns but was persuaded to wear them again by the monks in Geneva when he arrived there in 1579. When he published an attack on the University's distinguished member Antoine de La Faye, scandal ensued. Bruno fled once more and this time he arrived in France – first in Toulouse, where he obtained a doctorate in theology, before moving to Paris in 1581.

To make ends meet, Bruno set about lecturing and demonstrating his impressive memory techniques. It seems his memory was so good that a number of people believed he was a skilful magician. In particular, King Henry III of France wished to know if Bruno's memory was as good as rumour suggested or if it was merely some sort of conjuring trick. Bruno left the court of Henry III in 1583, living in England until 1585. During this time, he published his most influential works.

Bruno had accepted the Copernican model of the Solar System, but his own views extended much further. He suggested that the stars in the sky were other suns just like our own. Moreover, he believed the Universe was infinite, and the countless stars had planets orbiting them, and on these worlds he believed there resided intelligent beings. He believed that this increased the power of God; there was nothing special about the Earth, it was just one example of Creation, one note in the celestial symphony of life. This argument was to become known as the 'plurality of worlds'.

Bruno's work was not well received. Many people objected to the ideas of Copernicus simply because they could not conceive of an Earth in motion about the Sun. Bruno defended his works and was critical, sarcastic and plain rude to those critical of him. He appears to have lost the favour and support of many of his friends and, in October 1585, he returned once more to France. He did not remain there for long, since his support for Copernicus and his plurality of worlds ideology were unpopular. After an attack on the mathematician Fabrizio Mordente, Bruno fled France once more and made his way to Germany.

In 1591, Bruno received an invitation from Giovanni Mocenigo to go and stay with him in Italy. Mocenigo wanted to learn Bruno's memory techniques. Bruno must have been aware of the general feeling about him in Italy – the country which was the very heart of the Inquisition – but he believed that the Inquisition had lost some of its momentum, and he decided to leave Germany for Italy. He was quite wrong.

The relationship between Mocenigo and Bruno became acrimonious, and Mocenigo denounced Bruno to the Venetian Inquisition, which arrested Bruno on 22 May 1592. Bruno was skilful in defending himself and his works, but it did no good and he was transferred to Rome in 1593. Bruno was put on trial for Heresy and Crimes against God. Amongst the many charges, it seems that his theory of the plurality of worlds and his belief in the heliocentric model of the Solar System were the ones that offended the Church the most. It was at this point that the Church was realising that a Sun-centred Solar System seemed to be in contradiction to the accepted Ptolemaic model. It may have been the case that Bruno's memory techniques were seen as magical powers, further proof that he was a heretic.

The trial lasted for seven years, and Cardinal Bellarmino, who oversaw it, ordered Bruno to recant all his ideas and beliefs. Bruno refused, and Bellarmino branded him a heretic – Bruno's beliefs were now a heresy to the Church. An indignant Bruno was burned at the stake in 1600. The pursuit of Copernican ideology had become a dangerous game to play.

THE NEW ASTRONOMY

Many astronomers still rejected the heliocentric model of the Solar System. What was needed was some confirmation that the theory had a firm basis in the truth. Fortunately, one of the most talented pre-telescopic astronomers was about to make his debut, a theorist obsessed with geometry and order who would galvanise observation and heliocentrism into a set of fundamental laws which would see astronomy escape from the clutches of superstition.

On 14 December 1546, the Danish aristocrat and astronomer Tycho Brahe was born. A few years later, he went to live with his uncle, who wanted the young Tycho to become a scholar. In April 1559, Tycho started at the University of Copenhagen. Primarily he studied law but he soon became interested in astronomy. He later studied at the University of Rostock, where he engaged in a duel with a fellow nobleman over who was the superior mathematician. The duel ended with Tycho losing the bridge of his nose. He had a replacement made, probably out of bronze and copper, although paintings show him adorned with a golden nose which, legend had it, he used to take off and polish whenever he was losing an argument.

Tycho read as much as he could and became familiar with the theories of Copernicus. Tycho, himself a religious man, could not quite bring himself to accept the new cosmology offered by Copernicus. However he did realise that the only way to understand the Solar System was to make careful and precise observations of the positions of the six bright planets.

When his uncle died in 1565, Tycho returned home to his father's castle. It seems his father wanted him to pursue a career in law but, when his father died in 1671, Tycho decided to dedicate himself to astronomy. He established an observatory at Herrevad Abbey and equipped it with the finest pre-telescopic instruments, many of which he constructed himself. It seems that Tycho not only had a precise eye for making measurements, he also had a precise hand for crafting them. Tycho used the following instruments:

Quadrants: These instruments were designed to measure the altitude of objects in the sky. Tycho later constructed a massive quadrant when he established an observatory in Hven.

Sextants: A sextant was used to measure both the altitude of objects in the sky and the distances separating them.

Parallactic instruments: These could be used to determine the distance of objects from the zenith – the point directly overhead, 90 degrees from the horizon.

In 1572, a bright supernova appeared suddenly in the constellation of Cassiopeia. This had a quite profound effect on Tycho. The Aristotelian theory stated that the Universe beyond the Solar System was immutable, never changing: it was a constant eternal sphere surrounding the Earth. How could it be that a new star should appear in the firmament above? Tycho made many observations of the nova, and found that the new star exhibited no parallax. Moreover, unlike the bright planets, it did not wander about the night sky along the ecliptic. All this suggested that the new star was a long way away from the Solar System.

By 1575, Tycho had intended to settle in the town of Basel. King Frederick II of Denmark very much wanted to keep Tycho in his country and in his employ. As an incentive, he offered Tycho the island of Hven as a place to establish a new observatory; just as important, Frederick offered the funds to pay for it. Hven was to become a major astronomical facility of its day and had more than a hundred people working there during its most productive years.

Further celestial manifestations were to attract Tycho's attention. In 1577, a bright comet made its appearance in the night skies and, ever the diligent observer, he was probably the first to observe it.

He measured its parallax, and showed that the comet must be well away from the Earth and not a part of the atmosphere, as had previously been maintained.

1588 saw the publication of the second volume of Tycho's great work, *The New Astronomy*. The second volume was concerned with the bright comet and was published first, as the first volume was concerned largely with his observations of the supernova, which he had yet to finish analysing. The second volume also contains Tycho's own model for the Solar System.

Tycho could not bring himself fully to convert to the Copernican model. It seemed to be a step too far to remove the Earth from the centre of Creation. As a compromise, he developed a geo-heliocentric model. In this unusual model, the Sun and Moon orbit the Earth, while the rest of the planets in the Solar System orbit the Sun.

Previous models of the Solar System all had crystal spheres supporting the planets in their orbits. Tycho deliberately removed them from his model. This may have been due to the presence of the bright comet which would have had to have smashed through these spheres as it passed through the Solar System.

In 1599, Tycho found a new patron, the Holy Roman Emperor Rudolf II. He moved to Prague and established an observatory in a castle at Benatky nad Jizerou. Tycho had by now amassed a vast quantity of observations of the positions of the planets. Moreover these observations were the finest made in the pre-telescopic era. Tycho wanted to convert his data into a working model of the Solar System, but he lacked the skills required to turn the numbers into a working theory. What he needed was a man who not only possessed a fine mathematical mind, but was also familiar with astronomy and the new cosmology of Copernicus.

THE COSMOGRAPHIC MYSTERY

Johannes Kepler was born on 27 December 1571 in the German state of Baden Wurttemberg. His family were not particularly wealthy, and his father soon abandoned them. Kepler's mother was a healer and dabbled with magic and herbal remedies – a dangerous thing to do in medieval Europe.

At six years of age, Kepler had his first major introduction to astronomy – he observed in the skies the great comet of 1577. Kepler and Tycho, unaware of each other in 1577, but destined to work together to change the course of astronomy, were observing the same comet. Perhaps for once the comet could be seen as a good omen – at least for celestial mechanics.

Kepler became fascinated with astronomy, and was already displaying an unusual mathematical talent. The lunar eclipse of 1580 seems to have strengthened his interests further. Perhaps Kepler would have become a great observational astronomer like Tycho, but a case of smallpox left him without much use of his hands, and affected his vision. If Kepler was going to make a contribution to astronomy, he would have to do so with his powerful mathematical talents.

In 1589, Kepler went to the Seminary at Maulbronn. Kepler was a Protestant, and he had a strong faith. To Kepler, God was the ultimate mathematician; the hand of God could be clearly seen in the clockwork mechanisms of the Solar System. After school, he attended the University of Tubingen where he studied theology and mathematics. His mathematical ability impressed many of his teachers and peers, and it seems when Kepler was introduced to the ideas of Copernicus he accepted them quite readily.

Kepler had intended to become a minister after he had finished university but it was not to be. Instead, at the age of 23, he was offered a teaching post at a Protestant school in the Austrian town of Graz. Kepler was to teach mathematics and astrology, since astronomy had yet to become a separate subject. He was, though, an appalling teacher. He would mumble incoherently as his mind wandered from one topic to another. He would frequently embark on random tangents and utter aloud half-finished thoughts. No doubt the most popular part of Kepler's lectures were the endings.

During his time at Graz, Kepler considered some important questions. He wondered why there were only six planets in the Solar System. (He did not know about Uranus and Neptune as they had yet to be discovered.) He also wondered why the planets were positioned the way they were. It was during another dreary lecture that Kepler was to make what he believed to be a profound observation. On the blackboard, Kepler had drawn a circle. Along its perimeter were all the constellations of the zodiac. Kepler drew within this circle a triangle. Inside this triangle, he drew another circle.

Kepler noticed that the relationship between the smaller circle inside the triangle and the outermost circle had the same relationship as the orbit of Jupiter did to the orbit of Saturn. Kepler wondered if other shapes might connect the orbits of the other planets. Could this perfect geometry reveal the mind of God, he wondered.

Kepler believed he had answered his question as to why there were only six planets – he believed it was because there were only five perfect solids:

He constructed a model whereby the five perfect solids carefully supported the orbit of the six planets. He published the details of this work in his first major piece of work, *The Cosmographic Mystery*, in 1596.

Although initially the model looked promising, in the end Kepler could never make it work. The five perfect solids did not agree with the planetary orbits as well as they should. But Kepler

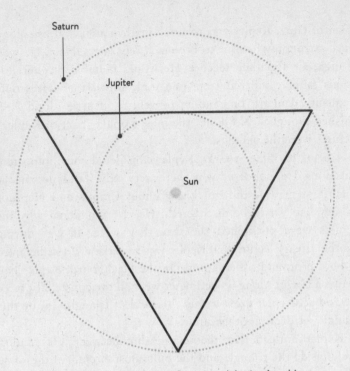

The relationship between Saturn and Jupiter's orbits,
as noted by Johannes Kepler.

had become obsessed with his idea and refused to abandon it. Indeed, rather than accept that the theory was wrong, he chose to believe that the observations of the planets were wrong – he believed that more accurate observations would provide agreement with his model of the Solar System. There was only one man in Europe with access to such observations, and that was Tycho Brahe.

In 1598, a surge in Catholicism made life in Graz become much more difficult. The archduke had made a vow to return Catholicism to his kingdom. Religious persecution started in earnest in Graz; those who would not convert were exiled. Kepler refused to

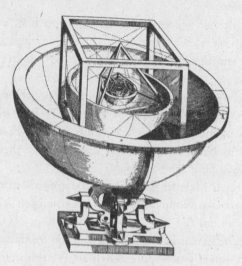

Kepler's model of the Solar System.

convert, so he and his family left. Fortunately, Tycho Brahe had invited Kepler to join him in Prague. The invitation seemed to offer the promise of quiet refuge, sanctuary from a world which was becoming increasingly hostile.

The man who Kepler met was not the Tycho of Kepler's expectations. Kepler was a rather quiet introverted character, in contrast to Tycho, who was loud and gregarious. Tycho liked to hold court and such occasions were wild and loud. One can picture the scene of a busy medieval banquet hall filled with merrymakers, music and copious amounts of rich food and wine. Tycho would be happily engaged with the revelry while Kepler glowered angrily in the corner, aloof and uncomfortable.

Kepler wanted to work on Tycho's observations, but Tycho was a proud man. Although he knew that he needed Kepler's help, he simply could not bear to hand over his life's work. So he revealed few of his observations to Kepler. This in turn frustrated and angered Kepler.

Tycho's lifestyle eventually caught up with him, and he died unexpectedly on 24 October 1601. In the delirium of his final hours, it is said that he repeated the same words over and over again: 'Let me not seem to have lived in vain.'

After Tycho's death, Kepler was appointed Imperial Mathematician and was duly given the task of turning Tycho's observations into a working model. Kepler was firmly of the opinion that the Copernican model was the correct one, and yet there was one planet which notably misbehaved.

If you watch Mars in the night sky it appears to move along the sky, stop, make a loop and then go back on itself. This is called retrograde motion.

This can be explained by considering the orbits of Earth and Mars. Mars moves much more slowly in its orbit than the Earth. As we approach Mars, it appears to move forward (i.e. eastwards) in the night sky. When we come to opposition (the time when the Sun, Earth and Mars lie on a straight line, with Mars opposite the Sun) the planet appears to stand still in our skies. Then, as we overtake the planet, it seems to us that the planet now moves backwards (i.e. westwards). We then move further away from Mars, leaving it behind, and it appears to us that it returns to its eastward motion in the sky.

Kepler examined Tycho's observations and found that the circular orbit was not in perfect agreement with Tycho's data. Put simply, if Mars were on a circular orbit, the mathematics would predict a position for Mars on a certain date and time which disagreed with the values which were actually measured by Tycho. Finally after many attempts, Kepler tried an elliptical orbit rather than a circle. To his surprise he discovered that the elliptical orbit worked perfectly; it gave accurate predictions for the position of Mars. Kepler had discovered his first law of

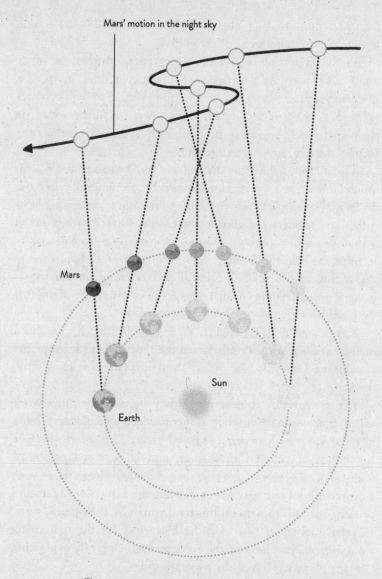

Mars' motion in the night sky

Mars

Sun

Earth

The apparent motion of Mars in the sky as it
and the Earth orbit the Sun.

planetary motion: all planets orbit the Sun in elliptical orbits, with the Sun at one focus of the ellipse.

As Kepler examined Tycho's observations further, he made another discovery. As a planet moves along in its orbit, its speed is not constant. If a planet has an elliptical orbit, there will be times when it is closer to the Sun, and times when it is further away. As the planet nears the Sun in its orbit, it moves quickly; when it is further away, it moves more slowly.

If we imagine a line from the centre of the Sun to a planet in orbit around it then, as it moves along in its orbit, the line will sweep out an area. When a planet moving close to the Sun travels along in its orbit, the imaginary line connecting it to the Sun will sweep out an area, A (see figure opposite). When that planet is far from the Sun in its orbit, the planet will sweep out a different area, B. Kepler found that during the course of an orbit, all planets sweep out equal areas in equal times. That means that it takes the same amount of time for the planet to sweep out area A as it does area B.

Kepler's third law is slightly more complicated. It provides a mathematical relationship between the time it takes for a planet to orbit the Sun (called its 'period') and the planet's distance from the Sun.

Kepler's laws of planetary motion were published in a work called *Epitome Astronomiae Copernicanae*, which was printed in three volumes between 1617 and 1621. They mark the true beginnings of celestial mechanics, and later, when combined with Newton's laws of gravity, they provided astronomers with a powerful tool not only for calculating the orbits of planets and comets, but also to help establish how massive the planets were. Celestial mechanics provided the blueprints for the underlying clockwork mechanisms of the Solar System, elegantly explaining why the planets behaved the way they did.

Kepler himself never really accepted his own laws. He was unable to abandon his beloved idea of the five perfect solids

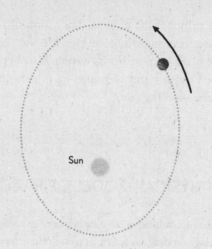

Kepler's first law: all planets orbit the Sun in elliptical orbits.

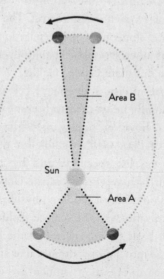

Kepler's second law: planets sweep out equal areas in equal times,
moving faster when closer to the Sun.

connecting the orbits of the planets. The vision of such a majestic structure proved too compelling for him. He was, of course, never able to make it work because it was wrong. Had Kepler lived to see the discovery of Uranus and Neptune, he might have finally been persuaded to abandon the theory.

A PHYSICAL MODEL OF REALITY

Although the firm foundations of celestial mechanics were established, there was still a great deal of doubt as to whether they were a correct physical model of the Solar System or if, instead, they were simply an accurate mathematical tool with no physical bearing in reality. What was needed now were further observations that Kepler's laws, based on a heliocentric Solar System, were correct. The new theory needed a powerful advocate, someone whose reputation would ensure the new cosmology was taken seriously. Thankfully, the last few strands all came together in one man: Galileo Galilei.

Galileo was born in Pisa, Italy, on 15 February 1564. He was a Catholic, and initially he wanted to become a priest. His father had different ideas, however, and wanted him to study medicine at the University of Pisa. Galileo studied a number of subjects, but mathematics was his specialty, and by 1591 he had been appointed the chair of mathematics at the University. Galileo can be considered the godfather of modern physics: he made a number of important contributions to the science including an early form of relativity whose basic premise was that the laws of physics are the same for anyone moving in a straight line at a constant speed.

Galileo had heard of an instrument which could magnify distant objects. This instrument, initially known as an 'optical trunke', was essentially a tube which had a larger primary lens at one end, and a smaller lens at the other. Light from distant objects was magnified by the primary lens, and focused by the secondary. In the summer of 1609, Galileo constructed his own; the era of the telescope had finally arrived.

On the night of 7 January 1610, Galileo Galilei turned his small telescope towards a bright point of light known as Jupiter. His telescope was no more powerful than a pair of 10×50 binoculars, but it was enough to change the course of human history. Instead of a solitary body, Galileo observed three bright objects in close attendance. Galileo watched as, night after night, the 'stars' – four in all – changed positions. One seemed to disappear, only to reappear later. Galileo was certain that these four bright objects were satellites of Jupiter. Their disappearance and subsequent reappearance he correctly deduced were due to the moons passing to and fro in their orbits, passing in front of and behind the planet.

This was the final clinching proof that the heliocentric model was correct, for if the Earth was supposed to be at the centre of all things, as stated by the geocentric model, then those moons should be orbiting the Earth rather than Jupiter.

Galileo also observed the planet Venus. As he watched, he noticed the planet exhibited all of the phases of the Moon – new, crescent, half and full. Only the heliocentric model could explain this as, in the old Ptolemaic model, Venus could never show a new phase, only gibbous and full.

The observations made by Galileo showed that celestial mechanics was not just a mathematical convenience, it was a physical model of reality. The Solar System was constructed with the Sun at the centre, and the planets orbiting it in elliptical orbits. Finally, the old Ptolemaic theory could be put to rest. It took a long time for the Church to accept this, however. Galileo defended heliocentrism; indeed, he saw no reason why it should contradict the Bible. The Church did not agree, and he was tried and found guilty of heresy. Although he was not executed, he was forced to live under house arrest for the rest of his life. He did receive visitors and continued to support the new science of celestial mechanics.

Once the telescope had been invented, it did not take long before it became popular. The new astronomers were not just armed with optical instruments, they were also armed with a powerful mathematical theory which brought order and understanding to the Solar System. The foundations established by Copernicus and others would allow humans to embark on a question to understand the laws of nature of the Universe, a quest that is still ongoing today.

* 4 *

The Sun

The Sun is incredibly important to our life – and in fact almost all other life – on Earth. Almost everything on the surface of the Earth is governed by the Sun – from providing a source of heat and light, to driving the temperature variations in the atmosphere that give rise to the weather and seasons. Quite simply, without the Sun, life wouldn't exist. Perhaps it is appropriate that for most of human history it has been revered and worshipped by almost all cultures at one point or another.

We've already seen in previous chapters that our view of the importance of the Sun in terms of the Universe as a whole has changed over time. First it was promoted to being the centre of the Universe, with the Earth and other planets orbiting it, and the distant stars being fixed in the sky. As far back as the 16th century, astronomers and philosophers were theorising that the other stars in the sky were much like our own Sun, but much more distant. Giordano Bruno and the early supporters of heliocentrism really were centuries ahead of their time. Proof of Bruno's ideas about the Sun and stars would take several hundred years, with the development of spectroscopy (see p.78).

SUNSPOTS

Although proper scientific study of the Sun didn't really begin until the first use of the telescope in 1609, people had been viewing the Sun for more than two thousand years, either with the naked eye (something that is not recommended) or by projecting it through tiny holes as in a pinhole camera or camera obscura.

These early observers noticed dark patches on the Sun's surface, and these became known as sunspots. Some are so large, at several tens of thousands of kilometres across, that they can be seen with the naked eye – though as we've said that is certainly not recommended without a fully certified solar filter. Ancient Chinese and Greek astronomers reported seeing them several hundred years BC. More detailed measurements began with the use of the telescope for Solar observations in the early 17th century. The sunspots appear black, because they are somewhat cooler than the rest of the Sun's surface. They are typically roughly 25 per cent cooler than the surrounding regions of the Sun, which lie at 5,800 Kelvin (5,500 Celsius), and so give out around a third as much energy. They only appear dark compared with the rest of the Sun – if they could be seen in isolation they would be brighter than an arc welder.

These spots were seen to move across the surface from day to day, and provided a way of measuring the rotation of the Sun. The Sun spins on its axis, just as the planets do, but it is not a solid body. While the equator spins once every 26 days, or thereabouts, the poles take around 38 days to make a full rotation. The sunspots appear to moves over the Sun's surface as it rotates, though they also appear and disappear with time, and even change shape and size.

This variation in the Sun's rotation causes its magnetic field to get wrapped around and tangled up. The sunspots are caused by concentrations of the Sun's magnetic field just below the surface, which inhibits the natural convection and so slows the flow of

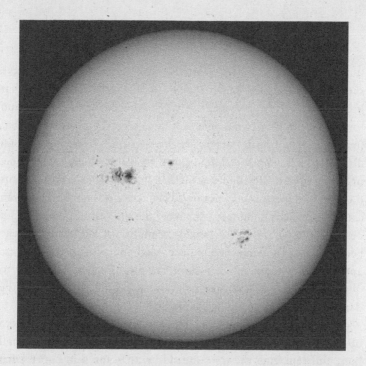

Sunspots seen by the SOHO satellite.

heat from the interior to the surface. Modern measurements have shown that sunspots, being magnetic phenomena, have north and south poles, and can give clues as to the internal magnetic structure of the Sun.

As sunspots appear and disappear, their number changes. Telescopic observations over the centuries have shown that the number of sunspots increases and decreases in a cycle with a period of approximately 11 years. This 'solar cycle' is thought to be due to the cycle of activity deep in the interior of the Sun. The length of the individual cycles varies, and there are longer timescale variations of hundreds of thousands of years, which affect both the number of sunspots and the Sun's activity and energy output.

OBSERVING THE SUN

It is possible, and indeed relatively straightforward, to observe sunspots on the Sun, but it is very important to remember that it can also be very dangerous. Anyone who has stared at the Sun for more than a brief moment will have seen the black spots that (normally temporarily) appear in their vision. Observing the Sun directly with any optical equipment, such as binoculars, a telescope, or even a camera viewfinder, is incredibly dangerous, as these focus the light and increase its intensity. The increased light and heat can cause permanent blindness, even after the briefest of glimpses. But despite these warnings we would certainly not like to discourage people from observing the Sun; provided you are careful, and use appropriate equipment, then there are a number of safe ways to observe the Sun and sunspots without exposing your eyes to the dangerous glare of magnified sunlight.

The simplest method is to use a pinhole camera, or camera obscura. This involves putting a tiny pinhole in one side of a small box, and replacing the opposite side with some thin paper, such as tracing paper. When the hole is pointed at the Sun, an image will appear on the paper opposite, which should be visible when viewed from behind. The further the paper screen is from the pinhole, the larger the image, but the fainter it will be as well. This essentially gives an unmagnified view of the Sun, but if you have a small telescope or binoculars it is possible to get a somewhat enlarged image.

You can use a small telescope or binoculars to project an image of the Sun onto a piece of white paper or card. The projected image can be made surprisingly large, and the view can be improved by surrounding the telescope or binoculars with a large piece of card, shadowing the projection screen from ambient sunlight. This method is safe as it removes the possibility of looking at the Sun directly through optical equipment, but there are some important

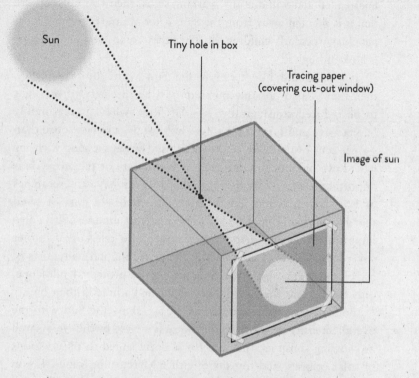

Sun

Tiny hole in box

Tracing paper
(covering cut-out window)

Image of sun

A camera obscura, or pinhole camera. An image of the Sun is
projected through a small hole onto a screen behind.

caveats. A reflecting telescope is unsuitable for projection, as
the high intensity of the light can cause the secondary mirror to
shatter. In addition, some telescopes, such those of the Schmidt-
Cassegrain design, are a sealed unit and the heat generated by the
sunlight passing through the telescope tube can cause it to explode.
So you must use a small refracting telescope, or binoculars, but
be aware that it is still possible to damage the eyepieces with the
projected sunlight, especially if they are made of plastic as many
of the cheaper ones are. Finally, if you leave your projecting set-up

unattended and without any warning signs, then you must ensure that it is pointed away from the Sun. After all, the first thing most people, particularly children, do when they come across a telescope is look through it.

If you would like to view the Sun with your own eyes, then that is also possible but requires the intensity of the light to be reduced considerably. A solar filter, which is reasonably inexpensive and fairly easy to get hold of, blocks out more than 99 per cent of the light, and can be used when viewing with the naked eye or optical aides such as binoculars or telescope. It is important to make sure that the filter does not have any scratches or holes in, and that it is securely attached – a gust of wind that blows it off could result in permanent blindness. It is also important to ensure that the viewfinder of a telescope is either covered, or removed completely. There are numerous reports of astronomers observing the Sun through a protected telescope, only to find that the viewfinder has burned a hole in their hair.

Both of the methods described above show the Sun with the overall intensity cut down by the time it is viewed by the eye. As well as showing sunspots, they are particularly suited to observations of solar eclipses, and transits of either Mercury or Venus. If you are in a position to view any of these events, then it is well worth preparing in advance. Try out your chosen observation method in the weeks leading up to the event, to ensure that you can do it safely and reliably.

'HYDROGEN ALPHA' FILTERS

Another popular method for observing the Sun is to use a filter that only lets through a very narrow range of wavelengths, or colours. The most common is a 'hydrogen alpha' filter, which shows the light given off by hot hydrogen gas – which makes up the vast majority of the Sun. By cutting out all the other colours, the intensity is reduced enough to make it safe to view with the eye. These filters can be purchased in a mounting suitable for attachment to a telescope, but are much, much more expensive than the white light solar filter mentioned above.

A better way of using these filters is to buy a bespoke solar telescope, which has all the filters fitted and is completely safe to use, with none of the safety concerns mentioned above. They are more expensive than a standard telescope, but the smallest versions, with main apertures a couple of inches across, retail at a few hundred pounds.

So why use these filters and telescopes, rather than the cheaper white light alternatives? Well, they give a very different view of the Sun. Since they only observe light emitted by hot hydrogen gas, and exclude the glow of the Sun, the structure in the surface of the Sun can be clearly seen. Viewed in this way, the Sun does not appear completely uniform, as convection below the surface causes the gas to rise and fall in convection cells, giving it a granular appearance.

But the real bonus is what can be seen just off the Sun's surface. Around the edge of the Sun are often loops of glowing material. These are prominences – loops of gas lifted off the Sun's surface by loops of magnetic fields – and can be several times the size of the Earth. Over the course of days, hours, or even minutes, these prominences can grow and shrink, rising above the Sun's surface and falling back down again. They are more common when the Sun is in a more active phase of its cycle as, like sunspots, they are caused by concentrations of magnetic fields.

THE CORONA

What we have referred to here as the 'surface' of the Sun is technically known as the photosphere. This is the region from where the vast majority of the Sun's light is emitted. But it is far from the outer limits of the star. After all, the Sun has no hard edge and so no solid surface, with the density of its material gradually falling away. In fact, as material from the Sun is continually flowing outwards (more on that later), it could be considered that the Sun is many billions of miles in radius, with its outer regions filling the Solar System and the planets gliding smoothly through it.

Beyond the photosphere is the corona, which could be considered to be the Sun's atmosphere. The corona does give off light, but that is all but completely washed out by the glare of the brilliant photosphere. It is only by blocking out the main glare of the Sun that the corona can be seen. Many satellites and telescopes achieve this by using a physical disk to block the light, but here on Earth we are fortunate to have a natural solution: solar eclipses.

By what appears to be happy coincidence, and nothing more, the Sun and Moon are the appropriate sizes and distances to appear almost exactly the same size in the sky. This means that on the rare occasions when the Moon passes directly in front of the Sun it obscures the main disc almost perfectly, blocking the light from the photosphere. Without the main glare of the Sun, the corona can be seen, making solar eclipses some of the most magical sights to behold.

The corona appears to extend away from the top and bottom of the Sun in a fan-shape, with arcs that trace out the Sun's magnetic field. The Latin name 'corona', meaning crown, derives originally from the Greek word κορώνη meaning wreath, and was given because of its appearance during solar eclipses. The glow seen from the corona is caused by gas in the Sun's atmosphere, which is at temperatures of millions of degrees. As well as scattering the 'normal' sunlight originating from the photosphere, this gas also emits its own glow. Although the gas in the corona is incredibly hot, there is very little of it compared with the main bulk of the Sun, and so it does not give off anywhere near as much light or heat. The corona is best seen not in visible light, but in ultraviolet light and X-rays, and it is these wavelengths that give us most clues about the structure and composition of the corona.

The cause of the corona's high temperature is not known for certain, but it is very likely to be related to the Sun's magnetic field. We normally experience magnetic fields that are caused by magnetic materials, and are commonly used to stick important notes to fridge doors. But they can also be caused by flows of electrically charged materials. In the heart of the Sun, the flow of the hot, dense ionised gas involves electrically charged materials moving around, and this causes a magnetic field to be generated. The Sun's magnetic field has north and south poles, just like a bar magnet used in a school laboratory, and extends from deep in its interior out into deep space. In fact, it doesn't end until it reaches the interstellar magnetic field, which is caused by the motion of material on massive scales within the Galaxy.

The interaction of the solar wind, which is constantly flowing outwards from the Sun, with the magnetic field threading through the corona, seems to allow huge amounts of energy to be deposited, which causes it to be heated to temperatures of millions of degrees – much hotter than the material on the visible surface of the Sun, which is only at a measly five and a half thousand degrees. Most of the material in the corona is made of charged particles – typically either electrons or protons, which are some of the main constituents of atoms. These charged particles follow the Sun's magnetic field close to its surface, and create staggeringly beautiful loops of ionised gas that appear to leap out of the photosphere into the corona, and then fall back again. But some of the particles are moving fast enough to break free of the Sun's gravitational pull and they continue flowing outwards, forming a 'solar wind'. In this way the Sun loses weight at a rate of several billion tonnes per hour – though there's plenty left and only a minuscule fraction of the Sun's mass has been lost in this way.

SOLAR FLARES, STORMS AND THE AURORA

The solar wind flows continuously, though it varies in strength and speed depending on the state of the magnetic field. Typically, it has a temperature in the region of a million degrees, and is made of particles travelling at several hundred kilometres per second (around a million miles per hour). The largest variations in the solar wind are normally due to the solar cycle, but there are sometimes events on much shorter timescales. These often occur in locations where the magnetic field is particularly concentrated, such as over sunspots.

When it is particularly tightly concentrated, small regions of the Sun's magnetic field can collapse in on themselves, releasing huge amounts of energy. Much of this energy goes into a flash of light called a solar flare, which appears as a brief brightening of a small patch of the Sun that lasts for a few minutes. The brightest flares can be easily seen with modest equipment, and the first observations were made in 1859, independently, by Richard Carrington and Richard Hodgson. As well as a flash of light, a magnetic disturbance was simultaneously measured by the instruments at Kew Observatory.

When these small regions of the Sun's magnetic field collapse, it is not uncommon for a significant amount of material from the corona to be flung outwards into the Solar System. These dramatic events are called coronal mass ejections, or solar storms, and consist of a billion tonnes of material moving at a million miles an hour. Their relationship with solar flares and magnetic phenomena on the Sun has been established through a long series of observations over many decades, stretching back to Carrington and Hodgson's observations in the mid-19th century. Until recently, however, we have only been able to monitor the storms that hit the Earth. While they do no lasting, physical damage to the Earth, they cause magnetic and electrical disturbances both in the atmosphere and down on the ground.

The most well-known of these disturbances take the form of eerie glows high in the atmosphere near the Earth's poles, called the aurora. The green and red glows of the aurora are emitted by energetic charged particles colliding with the atoms in the Earth's atmosphere and causing them to give off light of distinct colours. The charged particles flow along the Earth's magnetic field, which passes through the atmosphere in two oval-shaped regions (called auroral ovals) roughly centred on the Earth's magnetic poles, and rarely far from the Arctic and Antarctic circles. This localisation is what gives the aurora their more common name: the northern lights – or alternatively the southern lights in the southern hemisphere.

The particles that create the aurora are coming in all the time, not just from the Sun but also from very high in the Earth's own atmosphere. This means that the aurora can be seen relatively frequently during the dark winter months, though normally from a limited number of places. When a solar storm approaches, two things happen to increase the activity. First of all, there are many more particles coming in which makes the lights more vivid. But these storms are magnetic, and as a wave of magnetic disturbance passes the Earth, it distorts the Earth's magnetic field. This can make the auroral ovals larger, and push the northern lights further towards the equator. It is not uncommon on such occasions for the aurora to be seen from northern parts of the UK – and even sometimes on the south coast and northern continental Europe when there's a particularly good display on.

The northern lights are one of the most magical and awe-inspiring sights that nature provides. The glow forms curtains of green and red that seem to hang in the atmosphere, slowly drifting and writhing in the sky. They are hard to predict, and of course can be obscured by cloud, but there are ways to ensure you get a good chance of seeing them. In general: head north (or south if you're viewing the southern lights). From Europe, that generally means Scandinavia or Iceland. This is the reason why some of the world's leading research facilities for the aurorae are located in Tromsø in northern Norway. The aurorae are also often visible in North America – both from the northern United States and Canada. If you do go, make sure you stay for at least a few days to increase your chances of clear weather. The aurorae are only visible at night, and so it's best to go in winter. If you head into the Arctic Circle, then you might even have permanent darkness. The other way of observing the aurorae is to take an aurora-spotting aeroplane flight. These generally fly in a large loop, heading north to see the aurorae and then circling back to land where they took off. These are normally above the clouds and have a good chance of seeing the aurorae, but there's unfortunately still no guarantee that you will do.

THE SUN FROM SPACE

As we have learned to appreciate the strong effect it can have on the Earth, the Sun has come under increasingly close scrutiny over recent decades, both from ground-based telescopes and from satellites. Whereas earlier telescopic observations monitored little more than the sunspots on the Sun's surface, today's telescopes allow impressive views, showing the surface in incredibly high resolution. And it is not just professional astronomers, as modern technology allows amateurs to make observations that rival many professional observatories. All one needs is a modest telescope, a suitable filter (whether doing white light or 'narrow-band' imaging), and a decent camera. Professionals, however, have a number of advantages, most of which are down to budget, and one of the most significant is the ability to launch satellites into space.

The Sun has been studied from space since the early years of the Space Age in the 1960s. Many of the satellites in use today are very long-lived, with some having been in operation since the 1990s. In that decade the Ulysses, Polar, Wind and ACE satellites were launched to study the solar wind and the Sun's magnetic field – with particular attention paid to their effect on the Earth. In order to study the Sun's surface, the SOHO and TRACE satellites were launched, with SOHO still operating in 2013 having launched in 1995. Throughout its mission, SOHO has provided a constant view of the Sun and the surrounding region.

The majority of these spacecraft observed either from low-Earth orbit or between the Earth and the Sun, and so all saw the same face of the Sun. The exception was Ulysses, which started its mission in 1990 by flying out to the planet Jupiter. This might seem like going in the wrong direction, but by swinging round the giant planet it was able to loop up and pass over the Sun's south pole and back around over the north pole. By combining the data from Ulysses with that from other satellites, we could build up a model of the solar wind and the Sun's magnetic field – both of which proved to be more complex than previously thought. The one catch with Ulysses was that each orbit took 6 years to complete, as it had to journey all the way back out to Jupiter's orbit before coming back in. Over its 19-year mission, Ulysses only passed over the Sun's poles three times, though between these passes it did an awful lot of excellent scientific investigations of the Solar System away from the plane of the planets.

The first decade of the 21st century was when solar astronomers really strove for a three-dimensional view of the Sun and its effect on the Earth. The European Space Agency's (ESA's) Cluster mission consists of four satellites which orbit the Earth in close formation, travelling out to a distance of around 130,000 km. Together, these satellites explore the region where the Sun's magnetic field interacts with the Earth's.

A few years later, the STEREO mission launched, comprising two identical satellites which are slowly drifting away from our planet in opposite directions. One is moving slightly ahead of the Earth, and another lagging behind the Earth in its orbit, with the two spacecraft slowly getting further apart. These widely separated vantage points give us stereo eyes on the Sun, producing a three-dimensional view of the material travelling away. By viewing the Sun and its surroundings from multiple locations it is possible to track the material moving away from the Sun. It's not just professional astronomers who are doing this – through the solarstormwatch.org website, members of the public can help predict when coronal mass ejections are heading for the Earth – giving us advance warning of when we should prepare for any adverse effects.

The Solar Dynamics Observatory, launched by NASA in 2010, is one of the most recent and capable space-based solar observatories. Its main camera takes images of the Sun every ten seconds, with each one being better than high-definition TV resolution. The images, which are taken in both visible and ultraviolet light, show details on the Sun's photosphere, or surface, as well as in the corona, or atmosphere. Other instruments on board study the Sun's magnetic field, and the motion of the solar surface. Such measurements are crucial to understanding the interior of the Sun.

THE SUN'S INTERIOR

Of course we can only ever see the surface of the Sun. Below the photosphere it is made of plasma – gas which is so hot that the electrons have been stripped off the atoms – which prevents light from travelling easily. It doesn't stop it travelling altogether, but the light is bounced around by the electrons and ions, and prevents us from being able to see what's going on. But we do understand the physics of what occurs inside stars, partly thanks to theoretical models but also from a large degree of experimental work in Earth-based labs. We can also study the motion of the surface of the Sun and figure out what movements must be taking place underneath.

Although light doesn't travel easily through the plasma, sound waves do – it's very true that there can be no sound in a vacuum, but the Sun after all is not a vacuum. A sound wave is just a vibration, which travels through a substance, be that substance solid, liquid or gas. When the wave reaches the Sun's surface, it causes it to dip and swell slightly, and this motion is what is measured by telescopes on Earth and on spacecraft. The speed at which a sound wave travels depends on the properties of the substance it is travelling through. The Sun's density is much higher at its centre than on its surface, which makes the sound waves travel in curved lines. By watching the Sun very carefully it is possible to deduce what the interior structure is. The Sun appears to vibrate on large and small scales. On the largest scales, the sound waves are travelling from the centre of the Sun to the surface and back again, causing the Sun to ring like a bell (though not as tuneful). It takes about five minutes to make the round trip, so it would be a very, very low note if we tried to listen to it. This makes large regions of the Sun's surface move up and down at around 0.5 km/s (around 1000 miles per hour), though this is still only a tiny change compared to the 1.4 million km diameter of the Sun.

This whole field of research is similar in nature to the study of seismology, which uses the movement of the ground to track vibrations caused by earthquakes, and thereby study the interior of the Earth. The similar study of the Sun's interior even takes a similar name: helioseismology.

We know that just under the Sun's surface the plasma is seething and writhing due to a process called convection, whereby warm material rises and cooler material sinks. The same process transfers heat from the Earth's deep interior to its surface, helps a heated pan of water come to the boil, and even enables your radiators to heat your living room. Convection causes material to rise and fall in a relatively small loop called a convection cell. The tops of these convection cells can be seen on the Sun's surface and give it a granular appearance when viewed in high resolution – though each granule is somewhere around 1000 km wide.

To uncover the source of the Sun's energy, we have to go right to its heart. In the Sun's core, temperatures reach tens of millions of degrees, and strange things start to happen to the particles that are flying around. The incredible temperatures cause nuclei of hydrogen atoms to be smashed together at tremendous speeds, forming a larger particle – the nucleus of a helium atom. This process is called nuclear fusion and is occurring continuously, turning hydrogen into helium. Every second, 600 million tonnes of hydrogen are turned into 596 million tonnes of helium. The missing mass (4 million tonnes) is turned into energy and manifests itself as light – which we see as sunlight. So the Sun is continuously losing weight, but there's a lot more left to lose – weighing in at more than a thousand million million million million tonnes, the Sun has a long time before it runs out of hydrogen to burn.

SPACE WEATHER

The Sun is crucial for life on Earth, but it does have some less positive effects. The most dramatic of these are due to solar flares. If a solar flare and an associated coronal mass ejection occur on the Sun and are directed towards Earth, three things occur. Firstly there's a bright flash of light from the flare itself, both in terms of visible light and X-rays. These can be observed from Earth, but don't really have any adverse effect. What follows as part of the coronal mass ejection is a disturbance to the Sun's magnetic field and a sudden spike in the number of charged particles streaming from the Sun. Together, these events are usually referred to as a 'solar storm'.

Normally these have a limited effect, though that's largely due to decades of protecting sensitive devices. The magnetic phenomenon can affect some navigation systems, and the high number of charged particles can damage electronics. Satellites and other spacecraft have electronics that are protected from damage from high-energy particles, but during a solar storm there are many more passing by than normal. Many satellites are put into a sort of 'safety mode' until the storm abates. It is not uncommon for satellites to be knocked offline, either temporarily or permanently due to solar storms damaging their electronics. In March 2012 cameras on ESA's Venus Express spacecraft were rendered useless for a couple of days after a solar storm hit them.

But what of people in space? The effect of a strong solar storm on the human body is not dissimilar to radiation sickness – in fact it's essentially the same thing. Astronauts onboard the International Space Station have to retreat to a more heavily shielded part of the station when a storm passes. Luckily, we normally get a few days' warning thanks to satellites which are monitoring the Sun. Back in 1972, in the midst of the Apollo missions, a series of solar storms passed by the Earth and Moon. Luckily, they occurred between Apollo 16 and 17, so there were no astronauts in space or on the Moon. If there had been, they may have become ill and had to hastily return to Earth – though a strong storm could be fatal. This issue of space radiation is a major area of research in the field of human spaceflight, as adequate protection would be needed if we ever want to send people to Mars or beyond.

Down on planet Earth, we are protected from the worst of these effects by the Earth's atmosphere, but not always. The strong variations in magnetic fields can generate huge currents in wire and cables. In 1989, the power lines in Quebec, Canada, were knocked out by such a geomagnetic storm. Although serious damage and loss of life were avoided thanks to circuit-breakers tripping, the event did inspire power companies across North America and Europe to implement protection measures. The most powerful solar storm recorded hit Earth in 1859, just after Richard Carrington had made the first observation of a solar flare. The storm was powerful enough to cause telegraph systems in North America and Europe to short circuit, with some even catching fire. This was a very rare event, the sort of thing that only takes place a couple of times per millennium, but disruptive solar storms can occur several times per year when the Sun is at its most active.

The other major effect the Sun has on the Earth is our weather and climate. After all, the Sun is ultimately responsible for generating the vast majority of our heat and light. The energy from the Sun, combined with the Earth's rotation, is primarily what causes winds and ocean currents. The Sun's activity, and the rate at which it produces energy, varies on a roughly 11-year cycle. In sync with this cycle we observe a variation in the number of sunspots and solar flares. But there are also longer timescale variations in the Sun's activity and on the amount of energy it outputs.

Over the past few hundred years there have been numerous examples of the Sun appearing to behave in an unusual way. One of the most famous events such as this is the 'Maunder Minimum', a period in the late 17th and early 18th centuries when very few sunspots were observed. This period coincided with the coldest part of the Little Ice Age, which ran from the 16th to 19th centuries and during which northern Europe and North America experienced more cold winters than normal. It's not clear whether this slightly colder spell was experienced worldwide or only in some regions, and it's even less clear what the cause was. This is in part due to limited historical records (which are strongly biased towards northern Europe and North America), and because there are many other events which play a role. For example, more frequent volcanic eruptions are suspected of causing global cooling, essentially by making the atmosphere dirtier, and there were more of them in the Middle Ages than there have been recently.

There is fairly strong evidence that changes in the Sun's activity affect the Earth's atmosphere in minor ways, such as with the amounts of certain molecules in the atmosphere, and this allows us to infer the energy output and activity of the Sun over thousands of years. The conclusions so far are that this varies by small amounts over long timescales. But there is one thing that pretty much all solar scientists agree on: while the Sun can have an effect on the Earth's climate and weather (both globally and on more local scales), variations in the Sun's output alone cannot explain away the recent increases in global temperatures and other changes to the climate.

SPECTROSCOPY

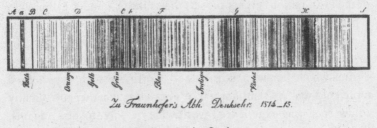

Fraunhofer lines in the Sun's spectrum.

The key to properly understanding the Sun is looking at its spectrum. Almost everyone is probably familiar with the spectrum, even if they don't know it – we see it every time a rainbow appears. The spectrum of visible light is simply the range of colours in the rainbow, from red through green, blue, violet and all the colours in between. The Sun normally appears white because it shines with all the colours of the rainbow, and when combined together they make white light. Spectroscopy is the study of the spectrum, which involves splitting the light from an object into its range of colours.

In 1802, the English chemist William Wollaston noted that the spectrum of the Sun had seven dark lines in it – narrow ranges of colour that were not emitted by the Sun. In 1814, the German optician Joseph von Fraunhofer invented the spectroscope, which combined the prism glass with a telescope, allowing careful study of the spectrum of an object. When Fraunhofer pointed his spectroscope at the Sun, he also saw dark lines, but rather than the seven seen by Wollaston he marked the positions of more than 500, and these lines are now given the name 'Fraunhofer lines'. Today, with modern instruments, astronomers have detected many thousands of these lines.

The origin of these dark lines remained a mystery until 1859, when the German scientists Gustav Kirchhoff (a physicist) and

Robert Bunsen (a chemist) used a spectroscope to examine the light given off when certain materials were burned. They discovered that different elements, or materials, emitted light only of very specific colours, called emission lines, and realised that many of them matched with Fraunhofer's dark lines in the solar spectrum. The conclusion was that the dark lines were caused by these same materials in the Sun's atmosphere, but instead of emitting light with those colours they were absorbing them, leaving only the colours in between.

COMPOSITION OF THE SUN AND STARS

The observations by Fraunhofer, combined with the work of Kirchhoff and Bunsen, allowed astronomers to study what the Sun was made of. Each element has a specific pattern of colours that it emits and absorbs, providing a spectral fingerprint. Combining these fingerprints with the observations of the Sun, it was established that the Sun is made of elements such as hydrogen, oxygen, sodium, calcium and iron. Further observations revealed that there were also some colours that appeared brighter, caused by material emitting light rather than absorbing it. Some emission lines seen in the solar spectrum had never been observed on Earth, which led to the conclusion that they were caused by an as yet unknown element. This element was named helium, after the Greek word for the Sun, *helios*.

Once it was established that the composition of the Sun could be measured, the focus of astronomers turned to observations of the stars. Being so much fainter, the spectra of the stars were harder to observe, but they still showed the characteristic pattern of lines similar to the Sun. Kirchhoff and Bunsen's results proved what had been theorised since the 16th century: that the stars are made of the same stuff as the Sun. Each star has a slightly different spectrum, since all stars have a slightly different temperature and composition.

* 5 *

Mercury

In Roman mythology, Mercury represented the fleet-footed messenger to the gods. This is very apt as Mercury, the closest planet to the Sun, spins around our central star in just under 88 days. Until recent times, little attention was paid to the planet. Being so close to the Sun means that the planet never ventures too far from it and it can be found low down in the glare of the early morning or evening.

The close proximity of Mercury to the Sun has hampered investigation of the planet. Indeed, until the MESSENGER spacecraft that arrived at Mercury 2008, the only other craft to have visited it was Mariner 10, which flew past the planet in 1974–1975. Mariner 10 had initially revealed a world not too unlike the Moon – a world devoid of atmosphere or air, a surface pock-marked with craters, one side baking in the heat of the scorching nearby Sun, the other side dark and cold. All in all, it seemed there was little of interest near the fiery heart of the Solar System.

The arrival of NASA's MESSENGER spacecraft has changed this view, and revealed a world more interesting than we realised. Mercury has also played its part in our understanding of gravity. For centuries after Newton, Mercury's orbit quietly whispered to astronomers that there was something amiss with Newtonian gravity, but it took a Swiss patent clerk called Albert Einstein to realise it. Mercury was instrumental in testing our current theory of gravity – general relativity – and today it is slowly revealing its secrets to us and painting an interesting but stark picture of the early Solar System.

ANCIENT OBSERVERS

With no cities, light pollution or high-rise buildings to obscure the horizon, our ancient ancestors would have noticed the dance of Mercury in the sky. The planet would appear in their evening sky and over a few weeks appear to fall rapidly towards the Sun. It would then appear in the morning sky and slowly climb away, always appearing to fail and fall back in towards the Sun once more. It must have seemed to the ancients that Mercury was under some strange supernatural force.

The ancient Chinese noticed the wandering of the planet Mercury. For them, the planet was associated with the north, and they called it Chen Xing (the hour star). Interestingly, it seems the planet has been associated with Wednesday by many different cultures. In Hindu mythology, for example, the planet is called Buddha and is said to preside over Wednesday, while the French word for Wednesday is *Mercredi*.

The Babylonians made many observations of the planet, the earliest of which date back to the first millennium BC. They too noticed its rapid movement in the sky. The early Greek astronomers named the planet Hermes (the Greek precursor to the Romans' messenger to the gods) when it was visible in the evening sky, and Apollo when it was in the morning sky. It was only later they realised that they were observing the same object.

EARLY TELESCOPIC OBSERVERS

Without a telescope, Mercury seems to be little more than a rapidly moving bright point of light. It is only when viewed through a telescope lens that it is transformed into something far more interesting.

No one person can lay claim to the invention of the telescope. You might as well ask who invented a stew. As is often the case in human history, there is an evolution of scientific ideas and techniques, many of which haphazardly come to fruition. It seems that the earliest recorded telescopes were of Dutch origin, and one of the earliest credited telescope makers was Hans Lippershey, a German-Dutch spectacle maker. These telescopes, known then as 'optical trunks' were very small refractors. A refractor has an objective lens at one end (called the primary). The light from an object being viewed through the telescope passes through the primary lens and down to an eyepiece where it is magnified further and focused.

The early telescopes did not have much magnifying power. They also suffered from an alarming amount of chromatic aberration. This effect is caused by the fact that white light is really composed of all the colours of the rainbow (you can show this by passing white light through a prism: out of the other end you get a rainbow). The different colours of light pass through the glass and refract slightly differently through the glass, and the end result means you get all sorts of spurious colour effects (called false colour) which look pretty but can make the image hard to focus.

Even with these shortcomings, the owners of these early telescopes made significant discoveries about the Solar System. This is not surprising since the Moon and planets are bright, and make an obvious target for investigation.

Galileo Galilei is of course the most famous early telescopic explorer, and yet he seems to have made no recorded observations

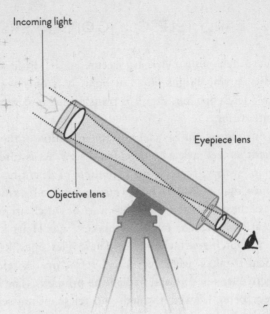

Incoming light

Eyepiece lens

Objective lens

A refracting telescope.

of the planet Mercury. Being closer to the Sun than the Earth, the planet shows phases like the Moon, and yet this fact was not discovered until long after the phases of Venus were observed by Galileo; such is the difficulty of observing the planet.

The first person to study Mercury in any great detail was Johann Hieronymus Schröter, who was born in Erfurt, Germany, on 30 August 1745. By trade, Schröter was a lawyer and by 1779 he achieved the appointment of Secretary to the Royal Chamber of George III in Hanover. Schröter was galvanised by the discovery of the planet Uranus by the famous Hanoverian astronomer William Herschel, who was then living in England (see Chapter 11). Schröter moved to the small town of Lilienthal, where he secured the job of chief magistrate, and quickly established an observatory there, equipped with first a

4.6-inch reflector, and then a 6.5-inch reflector. From here, he made extensive studies of the planets and the Moon. He made a number of interesting observations regarding Mercury.

On 26 March 1800, Schröter and his assistant, Karl Ludwig Harding, observed that the crescent of Mercury looked unusual. Rather than coming to a fine point, the southern part of the crescent looked distinctly blunt. (This effect has been observed by a number of astronomers right through to the 20th century.) Perhaps his most famous observation came on 17 March 1814, when Schröter observed a bright point of light in the southern polar region. When observing the Moon, it is common to see bright pinpoints of light in the shadows – these are the peaks of the lunar mountains catching the sunlight. Schröter wondered if he was seeing a similar effect, although a brief calculation revealed the mountain would need to be of an enormous height, and Schröter ruled this out.

Alas, tragedy was to befall Schröter, when the Napoleonic wars reached Lilienthal. Schröter's observatory was ransacked and destroyed along with all of his books and records. Schröter never recovered from the incident and, understandably, he did little astronomy after that.

The next astronomer to study Mercury in any great detail was Giovanni Schiaparelli. As we shall see later, the Italian astronomer would become famous for his observations of Mars, observations which would lead to much controversy. What is perhaps less well known is that Schiaparelli also made many observations of Mercury. Rather than observe the planet when it was a conspicuous morning or evening object, Schiaparelli chose to observe it during the daylight hours when it was high in the sky.

Schiaparelli was born on 14 March 1835. He decided at an early age to become an astronomer, and he was the director the of the Brera Observatory for over 40 years. The observatory was equipped with an 8.6-inch refractor and then later a 19-inch refractor, and it was from here that Schiaparelli made most of his important observations.

Over a period of time he made some 150 drawings of Mercury's surface, and it was from these drawings that he composed the first map of the planet. It seemed to Schiaparelli from his observations that Mercury always presented the same face to us, just as the Moon does. This led him to conclude that Mercury has a synchronous orbit – i.e. the length of a Mercurian day is as long as its year, some 88 days. As we shall see shortly, there are reasons why one could be fooled into thinking this was the case, and the synchronous rotation idea would hold sway until radar observations revealed differently in the 1960s.

The next map of Mercury came from the legendary visual observer Eugène Antoniadi, the finest visual planetary observer of his day. Antoniadi was of Greek origin, but he went to France to work with Camille Flammarion, who had established an observatory primarily to study Mars. Antoniadi used the great 33-inch refractor at Meudon, Paris, to make a number of drawings of Mercury, which he then turned into a map. Like Schiaparelli, he believed that Mercury had synchronous rotation. Antoniadi was a prominent member of the British Astronomical Association (and was one of the early Mars Section directors) and his map of Mercury was in vogue for many years.

OBSERVING MERCURY

Today, Mercury is seldom observed, but it does have a number of dedicated observers. Before we go on to look at how best to observe it, we shall need to know where to find it and for that we shall need to understand something about the orbit of this elusive world.

Astronomers refer to Mercury as an *inferior planet*. The name is not an aspersion cast upon it by the uninterested; rather it means

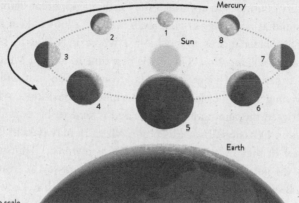

A simplified version of Mercury's orbit and phases as seen from Earth.

that the planet lies between the Earth and Sun. Under this definition, Venus is also an inferior planet, while the rest of the planets, Mars, Jupiter, Saturn and so on, are called superior planets.

All planets orbit the Sun in elliptical orbits, with the Sun at one focus of those orbits (see Chapter 3). As Mercury moves around in its orbit, its apparent diameter (i.e. the angular size of it as seen from the Earth) appears to change as the planet comes closer to us and then moves away. It also displays phases like the Moon, and Mercury's appearance in the sky is split into two types of elongation. If it is in the morning sky it is a western elongation; if it's in the evening sky it is an eastern elongation.

The image above shows a simplified representation of Mercury's orbit. At position 1, Mercury, the Sun and the Earth are all in a straight line. From our point of view, Mercury is very close to the Sun or sometimes behind it (though this does not happen very often, as Mercury's orbit is inclined slightly to ours). The planet is not visible from Earth – it is in *superior conjunction* with the Sun. As the planet moves out of conjunction, it appears in the evening sky, and a telescope will show it to be a waning gibbous (position 2).

As Mercury continues to approach the Earth, its apparent diameter increases and at position 3 it is at *dichotomy* and looks like a half-moon. The planet moves to position 4 where it is said to be at *greatest eastern elongation*. This is the best time to look at Mercury as, from our point of view, it is as far from the Sun is it will get in the evening sky. The planet then continues to position 5 where it lies between the Sun and the Earth and is said to be at *inferior conjunction*. Occasionally at this point, Mercury may pass in front of the face of the Sun in an event known as a *transit*. Transits of Mercury do not happen every time the planet passes in front of the Sun, because its orbit is tilted relative to the Earth's. They are, however, slightly more frequent than the better-known transits of Venus, with more than a dozen transits per century. The next transits will occur on 9 May 2016 and 11 November 2019; after that, the next transits will be in the 2030s.

After inferior conjunction, Mercury moves into the morning sky and embarks on a western elongation. The planet now looks like a waxing crescent. At position 6, Mercury is at greatest western elongation and this is the best time to catch the planet in the morning sky. It is furthest from the Sun, and rises many hours before the Sun. At position 7, Mercury is at dichotomy once more, and the planet continues to move away from us and is now a waxing gibbous as it reaches position 8. Finally the planet reaches superior conjunction at position 1 and the whole cycle starts once more.

If you live in a small town or a city (or indeed anywhere where the horizon is obstructed) you will find observing Mercury a challenge. As we have seen, the planet never strays far from the Sun, and there are principally two ways to observe it: either you can catch it in the early evening (or early morning, depending on which elongation it is on) or you can view it in the daytime.

When it is in the early morning or evening sky, it can be quite low down but bright enough to see easily. When low down, a view through a telescope is likely to be marred by poor seeing. ('Seeing' is a term astronomers use to mean how steady the air is.) When an object is low down in the sky, it is has to come through more of the Earth's atmosphere than if it was high up in the sky. The Earth's atmosphere is turbulent, and so the image in the telescope distorts and wobbles.

In fact, Schiaparelli had the right idea – the best time to observe Mercury is during the day, when the planet is high in the sky. A word of caution here, however: on no account go looking for the planet with binoculars or a viewfinder when the Sun is above the horizon. You may well inadvertently catch the Sun, and one second is enough to do a lifetime's worth of damage (see Chapter 4). The only safe ways to view Mercury in the day are to use a telescope fitted with a 'goto' mount or to use setting circles. You can obtain the right ascension and declination from the free software WINJUPOS and use the setting circles to dial up the position.

A 3-inch telescope should be enough to show you the phases of the planet. A view of Mercury through a larger telescope (6 inches or more) may reveal the presence of faint dusky markings on the surface. The planet has a rather orange colour, similar to, although not as striking as, the colour of Mars.

Filters can also be of help here. Light can be thought of as a wave, and different colours of light have waves of different length. An optical filter allows some wavelengths to pass but not others. Very often you will find telescope filters for visual use are marked with a number. This is called a Wratten number (named after the inventor Frederick Wratten) and each number corresponds to a particular colour: a W#11 is yellow green, a W#25A is red, a W#47 is violet, and so on. These filters bring out and enhance the contrast of planetary features.

THE ORBIT OF MERCURY:
A TESTING TIME FOR GRAVITY

At face value, Mercury's orbit looks reasonably straightforward. It orbits the Sun in an elliptical orbit alternating between east and western elongations. It should be the case that if we apply Newtonian gravity to Mercury's orbit, we can make predictions as to where the planet will be at any point in the future. When this was done, it was discovered that Mercury refused to behave properly. The planet seemed to stray from its predicted position. Even a number of refinements made to the planet's orbit were insufficient to explain why Mercury was still prone to wandering away from its predicted position. Initially, it was thought that this discrepancy might be the gravitational fingerprint of a planet even closer to the Sun than Mercury; however, once Einstein's theory of general relativity was established, it was discovered the true cause was a phenomenon known as *perihelic precession*.

If a planet orbits the Sun on an elliptical orbit, then it is clear that there will be a point in that orbit when it is closest to the Sun, and a point in that orbit when it is furthest away. The closest point a planet makes to the Sun in its orbit is called *perihelion*; the point furthest from the Sun is known as *aphelion*. The image opposite shows Mercury's orbit around the Sun, with the point of perihelion. So, imagine Mercury is at perihelion. It then moves around the Sun in its orbit, and some 88 days later it returns to perihelion again. Because of the strong gravitational field of the Sun, however, the position of perihelion has moved round ever so slightly, meaning that the entire elliptical orbit has shifted a small amount. So, as Mercury embarks on another orbit, this one is offset by a small amount from the previous one. As a result, each time Mercury passes through perihelion, the point of perihelion has shifted. Over a century, the total shift is around 43 arcseconds.

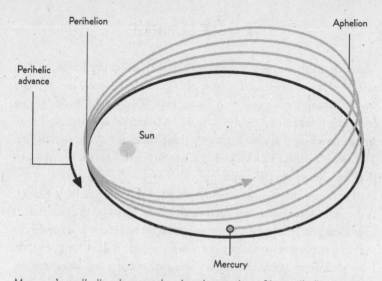

Mercury's perihelic advance, showing the motion of its perihelion position over time. The elliptical shape of its orbit is exaggerated for clarity.

This perihelic advance is not, however, sufficient to explain why the early telescopic observers thought that Mercury always presented the same face to Earth. This is thanks to a combination of factors. Mercury's synodic period (i.e. the times between successive phases) is about 116 days. So if Mercury appears as a half moon (50 per cent illuminated) today then it will be in exactly the same phase 116 days later. By coincidence, three synodic periods are almost the same length as one Earth year. Now, after three synodic periods, Mercury presents the same hemisphere to us and is well placed for observation, with all the same markings on show. It is a curious coincidence of planetary geometry and celestial mechanics, and we can now see that those early telescopic viewers were indeed viewing the same parts of Mercury's surface simply because they observed after three synodic periods. The cycle is broken after a number of years, but no one had made a long enough series of observations to realise this.

DOUBLE SUNRISE

Mercury orbits the Sun much more quickly than the Earth, taking just 88 days to go round, but it also spins on its axis much more slowly, rotating once every 58 Earth days. This slow rotation is due to strong tides caused by the Sun, and Mercury has ended up in the situation where it spins on its axis three times for every two orbits. The same process has locked the Moon's orbit to the Earth's, so that we always see the same face.

There are surprising consequences for this bizarre little planet. Although Mercury rotates on its axis every 58 days, the fact that it orbits the Sun in just 88 days means that on Mercury a 'solar day' – the time between two successive 'noons' – is 176 days long, exactly two Mercurian years. So if you were stargazing on Mercury, the stars would wheel round once every 58 Earth days, but the Sun would only rise once every 176 days.

Since Mercury's orbit is very elongated, with the planet moving between 46 million km and 70 million km from the Sun, the planet's speed increases and decreases over the course of its orbit. While the stars move overhead at a constant rate, the speed of the planet's motion means that the Sun's apparent motion can do some really weird things. From some locations, the Sun appears to move through the sky in the familiar way, from east to west, then seems to stop, move backwards for a little while, and then returns to its conventional motion for the rest of the day. From other places on Mercury's surface, the Sun rises and sets at the start and end of the day, as one might expect, but then it briefly pops up above the horizon again for a little while. There would be two sunrises per day.

MESSENGERS TO MERCURY

Although the Mariner 10 spacecraft visited Mercury in the 1970s, it was unable to go into orbit around the little planet. Getting into orbit around Mercury is surprisingly hard, as the spacecraft would need to slow down considerably, requiring a large amount of fuel. Mariner 10 used a gravitational slingshot around the planet Venus to slow it down and send it to Mercury, but the spacecraft was moving so quickly when it got there that it came back out to the orbit of Venus once again. It looped back in to Mercury another couple of times before the spacecraft ran out of fuel and remained in orbit around the Sun.

Mariner 10 made a number of key discoveries about Mercury, such as the presence of a large iron core, but was unable to map the entire surface. Part of the reason was Mariner 10's orbit, which took almost exactly twice as long as Mercury's and therefore brought it back to the same point in Mercury's year. And since Mercury's solar day is also two Mercurian years long, the same part of the surface was in sunlight each time Mariner 10 passed.

Although there have been some radar measurements made from Earth, we didn't get another close-up view of Mercury's surface until NASA's MESSENGER spacecraft swung by in 2008. The spacecraft performed three flybys before finally going into orbit in March 2011 – the first spacecraft ever to do so. Once there it unearthed a number of surprises. It discovered water vapour in Mercury's 'exosphere' – a very tenuous flow of particles escaping from the surface – and even saw evidence for water ice in the bottom of craters near the poles.

Some of the biggest questions about Mercury relate to its formation and history. Although the planet is 2,500 km in radius, it has an iron core which is thought to have a radius of almost 2,000 km and takes up nearly half the planet's volume – over twice the fraction of Earth occupied by its core. The large core was discovered by

Mariner 10, and for decades was thought to indicate that Mercury is the inner part of a much larger planet that experienced a massive collision in its early history. But the results from MESSENGER showed that this theory is probably wrong. Such a collision would have generated immense temperatures, melting much of the surface, and would have changed the chemical composition of the surface. Some elements, such as potassium, are more volatile than others, and evaporate away more easily at higher temperatures, and so a collision should have resulted in a greater loss of these elements than less volatile ones, such as thorium.

To test the theory, the surface composition of Mercury is compared with other objects in the Solar System for which we know the history. For example, the Moon, which almost certainly formed from a massive collision, has a fairly low abundance of potassium relative to thorium. The surface of Mercury, however, more closely resembles that of Mars, which did not experience such a massive collision in its early history.

While MESSENGER showed that Mercury did not experience a massive collision, Mercury's past has been far from uneventful. There is evidence of volcanic activity relatively recently in its past, perhaps as recently as a billion years ago. This was not in the form of traditional volcanoes, leaving high peaks, but rather the flooding of large areas of the surface by magma, possibly as a result of the mantle re-melting some time after its formation.

There are still many mysteries surrounding Mercury, not all of which will be answered by MESSENGER. But the BepiColombo satellite will launch in 2015, arriving in 2022. Named after Giuseppe Colombo, the Italian scientist who first derived the trajectory for travelling to Mercury, BepiColombo is a collaboration between the European Space Agency and JAXA, the Japanese space programme. It is ten times the size of the MESSENGER probe, carrying more sensitive instruments, and will help to answer many of the questions that MESSENGER has raised.

* 6 *

Venus

After the Sun, Venus is the brightest object in the sky when it is visible. Viewed against the dark azure blue of the early morning or late evening sky, the planet looks like a brilliant oil lamp floating serenely out at sea. Seen like this, it is not hard to understand why the Romans named the planet after their god of love and beauty.

Over the centuries, Venus has tantalised Earthbound astronomers. The planet's surface is permanently hidden behind a thick extensive cloud deck, and what might lie beneath these lemon-coloured clouds became a rich tapestry for both scientists and science fiction authors. Could it be that the planet was a tropical paradise where giant dinosaurs ruled with impunity? Perhaps the surface was covered in oil and would provide an almost unlimited amount of energy if it could be harvested? Others thought the planet might harbour an advanced civilisation, who occasionally visited Earth in their flying saucers. When there is no evidence to the contrary, speculation can converge on the most wild of ideas.

On 14 December 1962, the American Mariner 2 arrived at Venus and revealed not a tropical paradise, but a world bearing a close resemblance to hell. Thick poisonous clouds of carbon dioxide trap the Sun's heat, raising the surface temperature to a baking 425°C. They also create intense pressure – the atmospheric pressure on the surface is around 100 times greater than the Earth's at sea level. For many people, it was a shock to learn that our sister planet was as inhospitable as she was beautiful. Only those who had suggested a runaway greenhouse effect might have occurred saw the planet deliver the expected.

Although Venus was not as predicted, in many ways the planet has delivered greater mysteries than those just created by the minds of human beings. Today the planet's surface has been mapped, and there is still plenty for the amateur and professional astronomer to do.

ANCIENT OBSERVERS

Venus would have been visible to mankind's early ancestors. With nothing but trees to obscure the horizon, the early people of Earth could not have failed to have seen the dazzling orb lighting up their early morning and evening skies.

There are a number of ancient observations of the planet. The Venus tablet of Ammisaduqa is a Babylonian stone tablet from around 1581 BC. Here in cuneiform are observations the Babylonians made of Venus as it graced their morning and evening skies. It seems that the Babylonians were well aware that the morning star and the evening star were in fact the same object.

Both the Greeks and the Romans associated the planet with the goddess of beauty. The ancient Aboriginals thought Venus played a role in communicating with the souls of the dead. Both the Aztecs and the Mayans also observed Venus in their skies. Indeed, they invented a rather convoluted calendar based on the movements of Venus along with the Moon and the Sun.

TELESCOPIC OBSERVERS

Not surprisingly, the bright beacon of Venus made it an obvious study for the early telescopic explorers. One of the earliest was Galileo. On 11 December 1610, Galileo pointed his telescope (a small refractor which probably only magnified around ×40) towards the planet Venus. The telescope was very small, probably similar in power to a modern viewfinder, but it was enough to change the course of human history. Galileo observed that the bright orb of Venus exhibited phases, much as our own Moon does. The only explanation was that Venus was in orbit around the Sun, not the Earth as the Ptolemaic model suggested.

By 1639, the Ptolemaic model of the Solar System had been disposed of, and it was realised that both Mercury and Venus could pass in front of the Sun in an event known as a *transit*. These were the early days of the recently established *mathematical astronomy* pioneered by Kepler and Newton. Calculating the positions and times of the planets took a long time. Moreover, the data they had to work on was limited. It is not surprising that Kepler's Rudolphine Tables (a form of star catalogue and planetary table list) failed to predict a number of events. This became evident thanks to the work of two 17th-century British astronomers, Jeremiah Horrocks and William Crabtree.

Horrocks was the son of a watchmaker, and studied mathematics and astronomy in his spare time. After a visit to Cambridge he moved to Liverpool and began to acquire a library and the best instruments he could find. William Crabtree was also an amateur

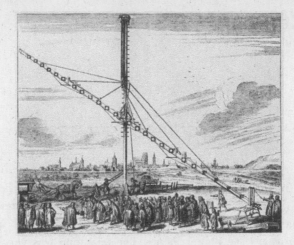

The 140 foot aerial telescope
used by Hevelius in 1673.

astronomer and mathematician in the town of Broughton, now part of Greater Manchester, and both Horrocks and Crabtree corresponded on astronomical matters of the day.

In October 1639, Horrocks made an impressive discovery (impressive as he would have had to have performed all the calculations by hand). He determined that transits of Venus occur in pairs, and each pair of events is separated by eight years, which Kepler had not recognised. Since the previous transit of Venus had been some eight years earlier, Horrocks realised that another was imminent. He wrote to Crabtree, and the two made plans to observe the transit from their homes.

Horrocks projected the Sun onto white paper and was able to observe the transit. Crabtree had bad luck – the weather at his location was rather poor – but he did get a number of observations and was able to estimate the diameter of Venus.

THE HIMALAYAS OF VENUS?

Once the telescope had been invented, it wasn't long before it became the main tool of observational astronomy. As better-quality optics gradually came into being, the Moon and other planets slowly began to reveal their innermost secrets, but Venus remained a stubbornly difficult target. It seemed that there were no prominent dark surface markings like those on Mars or Jupiter that could be used to make an estimate of the rotational period of the planet. Bright patches were seen on the surface in 1726 and 1727 by Francesco Bianchini, an Italian astronomer who studied Venus with a telescope some 100ft long. He observed a number of dark streaks and features on the planet and was so convinced of their reality that he made a map of them. Alas the features were largely fictitious and no doubt caused by the optics of his telescope.

In 1668, Sir Issac Newton had invented a new type of telescope. Rather than using a lens, Newton's design used a concave mirror instead. The light passes down the tube and hits the mirror where it is magnified. It then passes up to the secondary mirror where it is reflected into the eyepiece. Mirrors are much easier to make than glass lenses, and they don't suffer from chromatic aberration (see Chapter 5).

By 1788, Johann Hieronymus Schröter was observing Venus using a Newtonian reflector. His telescopes were the best available at the time (the mirrors were ground by Sir William Herschel, discoverer of Uranus; see Chapter 11). Many real features that we see on the planet today were logged. Schröter observed the elusive cloud patterns which keep a four-day rotation, and the extensions of the polar cusps which he explained correctly as the scattering of light. He also stumbled upon another phenomenon: it is always the case that, when in the evening sky, the observed phase of Venus is less than the predicted phase, while the situation is reversed when the planet is in the morning sky. This is

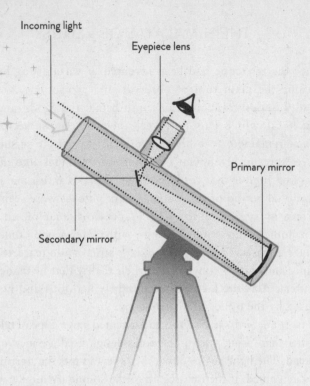

A reflecting telescope, as designed by Newton.

due to light being scattered by the thick Venusian atmosphere, and this effect is now called the *Schröter effect*.

Look at the Moon when it is around a half-moon, using binoculars or a small telescope. You will see that the terminator (the line which separates night and day) is not smooth. Rather, it appears slightly ragged due to the presence of mountains and craters. In the darkness, bright beads of light twinkle – these are lofty mountain peaks catching the sunlight. On the evening of 28 December 1789, Venus was at dichotomy and so 50 per cent illuminated from our point of view. As he examined Venus closely, Schröter noticed a couple of unusual things. For one thing the terminator wasn't a straight line; it seemed rather irregular. Even stranger, the southern cusp seemed to be somewhat blunted, and nearby a tiny point of light glittered mysteriously. He observed similar effects in 1790 and 1791 and came to believe that the underlying explanation for his observations was the presence of vast mountains on the planet. The Himalayas of Venus would have to be very high indeed to point up through the clouds.

Schröter wrote a paper about his observations and his belief in the mountains of Venus, but it was not well received. Other astronomers had failed to confirm his observations, and Herschel himself was deeply sceptical about the observations and the proposed Himalayas of Venus.

It would be easy to put Schröter's observations down to some sort of fault in his telescope or some equivalent explanation, but there is an interesting postscript to the story. Observations of bright points of light continued right up until the 20th century. Henry McEwen, director of the British Astronomical Association's Mercury and Venus section, observed 'sparkling stars' on the disc of Venus, and many observers have commented on the blunting of the cusps. It seems that, even now, Venus is determined to keep her secrets.

VENUS IN THE SKY

Like Mercury, Venus is an inferior planet and so lies between the Earth and the Sun. As you might expect, it has a similar orbit to Mercury and goes through the same phases.

We start when the planet is effectively behind the Sun, or at superior conjunction. The planet then moves into the evening sky and starts an eastern elongation. Early on, the planet is far away from us and so its apparent size is quite small. The phase is quite large and the planet is quite close to the Sun in the sky. When the planet is at dichotomy it appears to be half illuminated. Due to the Schröter effect, dichotomy can be a number of days before or after the predicted phase. After this, the planet appears as a large crescent. The phase continues to decrease but the planet's apparent diameter can get as large as 65 arcseconds as it approaches inferior conjunction. Here it is effectively between the Sun and the Earth. It is at this point that Venus occasionally passes over the surface of the Sun in the event called a transit. Although the inclination of Venus's orbit is small, it is enough to ensure that transits are rare. There was one in 2004, and the last one occurred in 2012, but the next one will not be until 11 December 2117.

After superior conjunction, the planet starts a western elongation and appears in the morning sky. The planet is moving away from us now, and so the phase starts to increase while the planet's apparent size starts to shrink. It becomes a crescent, then reaches dichotomy once more. The planet continues to move away, its phase increasing until it is quite close to the Sun in the morning sky. Venus continues to fall close to the Sun until it reaches superior conjunction once more and the whole cycle starts once more. Although Venus orbits the Sun in 225 days, its synodic cycle (the cycle of phases as seen from Earth) takes 18 months to complete.

OBSERVING VENUS

You might think that, being so bright, Venus is an easy object to observe telescopically, but appearances are once again deceptive. There are two problems associated with the planet. First, its brightness means that glare from the planet makes Venus hard to look at. This can be reduced considerably by observing the planet in the early evening sky or the daytime sky. Secondly, in the early morning or late

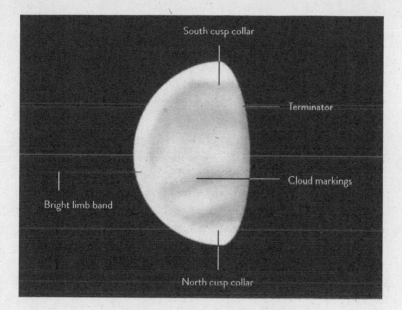

Image of Venus with common features.

evening the planet is low down and is affected by poor seeing. Most serious observers of the planet tend to observe the planet during the daytime as the planet is high in the sky and against the light blue sky. However, we must once more stress, never go looking for the planet with binoculars or a telescope when the Sun is above the horizon. The only safe way to find Venus in the day is to use setting circles and inputting the correct right ascension (RA) and declination (Dec).

A small telescope of 3 inches will show the phases of Venus very easily. Larger telescopes will reveal more subtle details. Don't be surprised if your first view of the planet is little more than a white disc. The markings on the Venusian disc are faint and elusive, and require some practice to see. As well as the faint dusky cloudy markings, after some time you should be able to make out the bright polar regions (discovered with a very small telescope in 1813 by Franz von Paula Gruithuisen), the bright limb band and dark terminator shading.

It is always interesting to watch the planet wax and wane. If you are lucky enough to observe the planet three days in a row, you might notice that the faint cloud patterns are not random but come and go, with a three-day rotation period. Another thing to watch out for is the Schröter effect. If you use a red filter you will see that the observed phase of the planet is not too far from the theoretical value; however, in a blue filter the phase of Venus is much less than the predicted value as the blue light is scattered more by the atmosphere of Venus, giving rise to the Schröter effect. Filters can be a real asset when looking at Venus. Optical filters allow us to observe different layers of the atmosphere. If you can make out markings on the disc, try experimenting with red, yellow and blue filters, and you will see the markings vary according to which wavelengths of light you are looking at.

AN UNRESOLVED MYSTERY: ASHEN LIGHT

There is one more phenomenon unique to Venus whose controversial existence is still debated today. Over the centuries, a number of telescopic observers have reported seeing a faint glow on the dark side of Venus. Normally the glow is a greyish-green colour and can be contained to a small region, or all of the night side of the planet. This phenomenon has become known as the Ashen Light. It was first reported by Johannes Riccioli in 1643, and a number of reliable observers have reported seeing it over the centuries, and as recently as 2010. To be seen, the phase of Venus must be small, which means that planet is large and low down and the glare from the thin crescent can be overwhelming. Some observers use a special eyepiece with an occulting bar that blocks out the bright crescent leaving just the dark side.

If the light is real, what could cause it? Rituals from Venutians themselves? It is unlikely to be because of the hot surface; the radiation (infrared) would be far beyond the range of the human eye. Another theory suggests that the phenomenon may be caused by rapid bursts of lightning which ionise parts of the atmosphere, causing it to glow. Yet another theory states that it is little more than illusion caused by the brilliant crescent being seen through our own turbulent atmosphere. Whatever the reason, keep an eye out when the planet's phase is small – you might be the next person to catch a glimpse of this puzzling phenomenon.

TRANSITS OF VENUS

If the orbits of the planets lined up perfectly, Venus would pass directly between the Earth and the Sun roughly once every 18 months, but the Solar System is far from perfect. The orbits of the planets are all tilted slightly relative to each other, with Venus's being inclined at around 3 degrees relative to the Earth's. As a result, from our point of view Venus almost always passes either just above or just below the Sun, and only rarely moves directly in front.

Of course, rare events do sometimes happen. Johannes Kepler's prediction in 1627 that Venus would pass in front of the Sun in 1631 proved correct, but the event took place during the night in Europe. Venus was therefore below the horizon, and so was not observed. Kepler also predicted that Venus would just miss the Sun eight years later, in 1639, but only because his calculated orbit was not quite right. Observations of a transit allowed the orbit to be refined and an accurate sequence of transits predicted. The eight-year interval was, it turns out, rather lucky, as the next transit was not for another 122 years, in 1761.

The 2012 transit of Venus, courtesy of Asadollah Ghamari.

We now know that Venus crosses in front of the Sun roughly twice per century. The specific alignment of the orbits means that transits occur in pairs, eight years apart – hence the transits in 1631 and 1639 – but that the sets of pairs are separated by over one hundred years. As such, mankind has only ever observed seven transits of Venus: the first observed event in 1639, a pair in 1761 and 1769, a pair in 1874 and 1882, and the most recent events in June 2004 and June 2012. From the UK, the 2004 event took place in the middle of a summer's day, and remarkably the weather held out over much of the British Isles. The 2012 event was less favourable from here, though, finishing just after the Sun rose, and on the day the country was largely shrouded in cloud. Since the next transit is not until 2117, we'll have to leave it to our descendants to observe that.

The reason the transit of Venus was not observed until the 17th century was that it is only easily visible with the aid of a telescope. Not only is the Sun difficult and dangerous to observe, but also Venus looks very small when compared with the Sun. As seen from Earth, Venus appears to be only around one-thirtieth of the Sun's diameter, and so the black silhouette it creates is very small indeed. However, while transits of Venus may not be as spectacular as solar eclipses, where the Moon obscures most – if not all – of the Sun's disc, they have been of scientific value over the centuries.

THE SIZE OF THE SOLAR SYSTEM

It was realised early on that transits of Venus could be used to measure how big the Solar System is. It may seem surprising, with all our knowledge of planetary orbits and our abilities to predict their positions, that this was not already known. But in the early 18th century, although the orbits of the planets were well understood, their distances from the Sun were only determined relative to each other. From Johannes Kepler's laws of planetary motion, the fact that Venus orbits the Sun in 225 days, compared with the Earth's 365, tells us that it is just under three-quarters of the distance from the Sun. But there is no easy way of knowing the actual distance, as astronomers can generally only measure the angular size of an astronomical object – that is to say the size it appears in the sky. To determine the distance to an object, its actual size needs to be measured, and the only object we knew the size of was the Earth.

The solution for the problem of measuring the size of the Solar System turned out to be the parallax (see Chapter 2). The key was to observe the transit of Venus from as many places around the world, and as widely spaced as possible. From each location, Venus appeared to cross the Sun in a slightly different position, with more northerly observers seeing Venus seeming to cross the Sun lower down than it did for more southerly observers. Similarly, those further east would see Venus start to cross the Sun very slightly earlier than those observers further to the west – though only by a matter of seconds. It is this latter effect that was the key to measuring the size of the Solar System as, in the 18th century, measurement of times was far more accurate than the direct measurement of positions. Most of the telescopes used were relatively modest, not least because they needed to be portable to get them to remote locations.

The observations of the 1761 and 1769 transits of Venus were massive international scientific endeavours, with observers travelling to Europe, North America, Russia, Norway and the South Pacific – the last being one of the stops on Captain Cook's great voyage of discovery. Using accurate clocks, these observers measured the precise times at which Venus started crossing the Sun's disc during the transits, and collated their measurements. There were a number of difficulties, ranging from optical defects, to cloudy weather, to blindness, with many fascinating stories about these trips that are sadly beyond the scope of this book. Despite these difficulties, through this collaborative effort the distance to the Sun was measured as being 93,726,900 miles – very close indeed to the currently accepted value of 92,955,807 miles (149,597,871 km).

This value was refined during the 1874 and 1882 transits, but by the time the 2004 transit came around much more accurate methods of measuring distances, such as radar, allowed the distance of Solar System objects to be measured much more reliably. In 2004, the European Southern Observatory organised a modern observing campaign called VT-2004, taking observations from over 1,500 people, including school students and amateur astronomers, and measuring the distance to the Sun within 11,000 km of the true value – an accuracy better than one part in ten thousand.

The 2004 and 2012 transits were largely used for different scientific purposes, regarding the atmosphere of Venus. As it passed in front of the Sun, the tiny amount of light absorbed by its atmosphere gave tell-tale signs as to the composition of the Venusian clouds. Since Venus is so small, the effect was tiny, and the astronomers were able to compare their results with the measurements by probes that had visited Venus. The reason for performing this seemingly academic exercise is to test methods and techniques, as these are the same as those used when studying the atmospheres of planets around other stars.

VENUS'S ROTATION

For millions of years, right until the mid-20th century, the clouds of Venus were all we could see of the planet. They are one of the reasons it is so very bright, as they reflect three-quarters of the sunlight falling on them – making Venus the most reflective planet in the Solar System. With almost completely featureless clouds, it was not even possible to study the rotation of the planet. That changed in the 1960s, when radar observations allowed measurements of the surface itself, including its rotation rate. These were initially somewhat confusing, providing unclear results, until it became clear that Venus was rotating very slowly. Venus takes 245 Earth days to spin on its axis, which is even longer than the 225 Earth days it takes to go around the Sun. Not only that, but it appeared to be spinning backwards – something called *retrograde rotation*. The vast majority of the planets in the Solar System spin in the same direction that they travel round the Sun – they could be thought of as rolling round, though that is obviously not what is actually going on. At some point, Venus appears to have been given a serious amount of backspin – though the origin of this is far from clear.

It might be useful here to think about why planets are spinning in the first place. They formed from a cloud of gas and dust, left over from when the Sun formed. This cloud was probably not spinning very quickly at first, with different parts moving in different directions, though there would have been a slight preference for rotation in one direction or another. As the cloud collapsed to form the Sun, the rotation increased. This is for the same reason that ballerinas and ice skaters spin faster and faster as they pull their arms in – a property of any rotating body called angular momentum, which can be loosely defined as the amount of spin. Angular momentum is hard to lose, and objects that move closer to the axis of rotation end up rotating around it even faster. This means that as the cloud of gas and dust collapsed it started spinning faster and faster. In the centre of the cloud the Sun was formed, but around it the rest of the gas and dust ended up forming a flat disc. Within that disc small grains of dust coagulated to form larger grains, and these stuck together to form pebble-sized objects. This process of building up continued, interrupted frequently by collisions that broke everything apart. But, in the end, planet-sized objects were made. As each bit of debris that ended up forming the planet was accreted into the main object, it carried with it some angular momentum, or spin. The effect of this accretion is to form a planet which is rotating in the same direction as it is orbiting, and the same is true of moons orbiting around planets.

Although it is hard for a planet to lose its angular momentum, it is not impossible. A massive object might have hit Venus in just the right way to start it spinning it in the wrong direction, although the chances of this are very small. For a start, it would have had to be lined up precisely, otherwise the planet might have ended up tipped on its side rather than spinning backwards. There are, however, more subtle effects that could be responsible. For example, strong tides make the rock move around within the object, generating friction and slowing the rotation. We experience tides on the Earth caused by the Moon. These don't only cause the oceans to rise and fall, but also create similar effects in the rock itself. Similarly, the Earth raises tides in the Moon, though since the Earth is much more massive the tides on the Moon are much stronger. Over billions of years, these tides have slowed the Moon's rotation so much that it spins on its axis at the same rate at which it orbits the Earth. The same forces have also slowed the Earth's rotation slightly, though the effect is smaller. Venus has no large moon, and so this seems unlikely at first, but there are also tides caused on the planets by the Sun. These solar tides are stronger on Venus than on Earth, but still unlikely to have been able to stop Venus rotating and start it spinning backwards.

It is possible, however, that Venus's thick atmosphere could be to blame. Although the planet itself is spinning very slowly, the atmosphere whips round in just four Earth days – sixty times faster than the planet itself. The Sun causes huge tides in the rapidly rotating atmosphere, and it has been theorised that friction with the surface could have slowed down the rotation of the planet itself. One piece of evidence that might support this comes from two of the recent spacecraft that have studied Venus: NASA's Magellan, which visited in the 1990s, and the European Space Agency's Venus Express, which arrived in 2006. Odd anomalies seen between the maps from these two spacecraft can be resolved if it is assumed that the planet's rotation period slowed down by six and a half minutes in the sixteen years between the two missions. That might not sound like a lot – it's just 2 parts in 100,000 – but if it can change that much within a couple of decades, it could change by a huge amount over the course of millions of years.

We know that events on a planet can have an impact on its rotation. Here on Earth, for example, earthquakes can alter the length of a day, but only by a tiny fraction of a second. Could it really be that the rotation of the atmosphere on Venus can slow the planet down, possibly even stopping it and spinning it the other way over millions or billions of years? The jury's still very much out, as there are problems with both this idea and the competing theory of a massive impact. The problem is that these kinds of effects are very hard to study – we don't have anything like them on Earth – and so they can only be modelled in computer simulations. And of course, like so many things in astronomy, we could be barking up the wrong tree altogether.

VENUSIAN VOLCANOES

For millions of years, all mankind has been able to see of Venus has been the tops of its clouds. The first few spacecraft to fly past in the 1960s and 1970s saw that the atmosphere was incredibly thick and dense, with the pressure on the surface being almost one hundred times that on Earth. This thick atmosphere, which is almost entirely composed of carbon dioxide, has given rise to a runaway greenhouse effect, raising the surface temperature to 500°C. The clouds in Venus's atmosphere are made of sulphur dioxide, and it is these that reflect most of the sunlight. Reactions between this sulphur dioxide and water vapour causes a rain of sulphuric acid, although the temperatures are so high that this evaporates before reaching the surface.

These conditions are highly incompatible with surface exploration, as probes have to endure crushing pressures and searing heat. Despite these challenges, Venus was a favourite target of both the Soviet and American space programmes in the 1960s and 1970s – being the next easiest body to reach after the Moon. There were a number of missions to study the atmosphere, with experiments suspended from

A synthesised image of Maat Mons, based on data from the Magellan spacecraft. The vertical scale has been exaggerated around 20 times.

either balloons or parachutes, though many of these were crushed during descent. The first successful landing was the Venera 7 probe in 1970. The record for the longest-lasting probe on Venus is Venera 13, which transmitted data for over two hours, even managing to send colour images, before succumbing to the harsh environment.

Given the inhospitable conditions both on the surface and in the atmosphere, the safest way to observe Venus is to stay in orbit, and several probes have provided incredible results. Given that the atmosphere is completely opaque, spacecraft cannot see the surface with conventional cameras, and have to map it using radar. The reflection of the radio waves gives information as to the topography (how high or low it is) and also the roughness (smoother surfaces reflect radio waves less than rougher surfaces). In the 1990s, NASA's Magellan spacecraft mapped almost the entire surface to a resolution of a couple of hundred metres. The radar maps from Magellan show that the surface is littered with volcanoes, and has been largely shaped by tectonic processes. But although there are hundreds of volcanoes on Venus, it is a matter of active debate as to whether any of them remain active, and it appears that volcanic activity all but ceased several hundred million years ago.

The age of the surface is calculated by studying craters caused by meteorites – those that make it through the thick atmosphere without burning up. While there are no craters less than a few kilometres in diameter (the objects that might make them are destroyed while coming through the atmosphere), the larger ones that do exist can be used to estimate the age of the surface. The number of craters on Venus is compared with the number seen on the Moon (for which we know the surface age much more accurately), and show that the entire surface is about the same age – between 300 and 600 million years old. In contrast, the surface of the Earth has a wide range of ages, with some regions being billions of years old and others forming right now. The constant movement and tectonic activity on the Earth also recycles the surface, gradually destroying any craters that have been formed as the continental plates crash together (albeit very slowly). On Venus, meanwhile, almost all the craters are in good condition, implying that the surface has been largely unaltered since they were formed. What this tells us is that for the past half a billion years or so Venus has had little or no tectonic activity. But, since none of the surface is older than that, we can't tell what happened before that – whether there was one single event that caused the whole surface to be melted and reformed over a relatively short period of time (100 million years or less), or whether there was a long period of constant volcanism and surface recycling that stopped around half a billion years ago. So what happened half a billion years ago to change the tectonic activity on Venus in such a dramatic way? And how (let alone why) do volcanoes stop erupting on a planet-wide scale? To understand this, we have to look at the other differences between the surfaces of Earth and Venus, such as the fact that there are no high mountain ranges on Venus, and the fact that Venusian volcanoes are scattered over the whole surface, rather than clustered along the boundaries of tectonic plates as terrestrial volcanoes are. All of these observations point to the same conclusion: tectonic activity on Venus today is very different from that on Earth. It would appear that it is all to do with how the rock behaves in the conditions that prevail on Venus.

First, let's briefly consider the situation on Earth. The surface that we see is the crust, a layer between 10 and 30 km thick made of a range of different types of rock, some formed in volcanoes and others laid down under rivers and oceans. The crust lies on top of a region called the mantle, which can be thought of as molten rock. Heat from the Earth's core generates convection in the mantle, which flows up, around and down in massive loops. This not only transfers heat to the surface, but also causes the crust to move around and be broken into large sections called tectonic plates. Some of these plates are old and relatively thick, forming the continents, while others are newer, thinner and denser, forming the ocean floors. As these plates move around, they can crash into each other (though only very slowly), either creating huge mountain ranges (such as the Himalayas) or forming chains of volcanoes as an oceanic plate sinks beneath a continental one (as is the case with the Andes mountain range). In the latter case, the molten rock is thrust back up to the surface, forming some of the most violent volcanic eruptions on the planet. Of course, if the Earth's crust is being destroyed in one place, it must be recreated somewhere else. As the tectonic plates move apart and fissures open up, molten rock rises to the surface and solidifies, forming new stretches of crust. This is taking place all along the mid-Atlantic ridge, as crust is constantly being created and moved away, leaving room for yet more crust to be formed. In some places the convection in the mantle creates a plume of hot, molten rock, which continues to rise to the surface in one spot. The Hawaiian island chain is such an example, with lava continuing to flow from the flanks of the volcanoes, forming some of the tallest mountains on Earth. There is one key ingredient which affects how these tectonic processes behave: water. It may seem odd, but small amounts of water exist in rocks both on the Earth's surface and beneath it – at levels of up to a few per cent. Many of the tectonic processes on Earth take place under or near to water, which becomes mixed with the molten rock. This makes it slightly less viscous (allowing it to flow and bend a little more easily) and easier to stretch, strain and break.

On Venus, the rocks have a similar composition to the rocks formed in volcanoes and along mid-ocean ridges, but are completely lacking in any water as Venus is an almost completely dry planet. This makes the crust several times thicker and also much stiffer and stronger, preventing it from breaking into plates which move around on top of the mantle. Without plates moving around, most of the tectonic activity we see on Earth simply can't occur on Venus. The volcanoes on Venus are analogous to those in Hawaii, formed over hot plumes within the mantle. But with such stiff, strong crust, most of these volcanoes seem to be dormant. But perhaps not all – there are hints from ESA's Venus Express spacecraft, which has been in orbit since 2006, that the surface around some of the hot spots is much younger, possibly less than a million years, which would mean that there is volcanism going on on Venus. In addition, Venus Express has detected lightning in the Venusian clouds. While lightning on Earth occurs in clouds made of water vapour, it shouldn't be possible in Venus's sulphur dioxide clouds. Instead, it is postulated that volcanic ash high in the atmosphere could cause lightning, though this ash has yet to be detected.

But whether there are active volcanoes on Venus today or not, we still haven't uncovered what happened half a billion years ago. One possibility is that conditions just beneath the planet's surface reached a critical point, causing lava to flow onto the surface all over the planet. After a few tens of millions of years things settled down again, but only after 80 per cent of the planet had been resurfaced. If that's the case, then perhaps such resurfacing has occurred before and will occur again.

Another possibility is that the lack of tectonic activity on Venus is tied in with its hellish climate. We know that it used to have much more water than it does now, probably flowing on its surface in rivers and oceans, as well as in its atmosphere. In fact, Venus probably used to be a much more pleasant place – possibly even Earth-like. As the atmosphere thickened and the surface heated up, these oceans evaporated into the atmosphere, leaving the surface completely dry. But the origin of the sulphur in the atmosphere, which is thought to have initiated the runaway greenhouse effect, probably came from volcanic eruptions in Venus's past. So it's possible that intense volcanic activity long ago in Venus's past caused an increase in greenhouse gases, leading to the disappearance of all the water, which in turn led to the cessation of volcanic activity.

Whatever the cause of the lack of tectonic activity, it may have had a significant impact on the interior of the planet, namely the generation of a magnetic field. On Earth, it is convection and rotation in the liquid outer core that generates the planetary magnetic field. Without a heat flow through the crust in the form of volcanic activity, it may not be possible for convection to take place in the liquid layers of Venus's core, and without that there can be no magnetic field generated. This is merely a hypothesis, as the interior of Venus is very hard to study. To find out more we would need very accurate measurements of the planet's rotation, preferably taken from the surface, or the placement of seismic experiments on the surface to study the way in which earthquakes travel through the planet's interior. Unfortunately, we have never been able to place any experiment on the surface of Venus for more than a few hours, and so its interior remains a mystery.

DESTINATION VENUS?

With crushing atmospheric pressures, searing surface temperatures, and the lack of a protective magnetic field, Venus certainly doesn't sound like a pleasant place to take a holiday. However, it's not all doom and gloom for future Venusian package holiday tours. For a start, Venus does have a weak magnetic field, but it is generated high above the planet's surface. Intense ultraviolet light from the Sun breaks apart the molecules in the upper atmosphere and strips the atoms of their electrons, forming something known as an *ionosphere*. This creates a shell of charged particles around the planets, which move around and generate a weak magnetic field. The same thing happens on Earth, but in Venus's thick atmosphere the effect is even greater.

This means that there is some protection from the harsh radiation environment of deep space, but there's no easy way round the high temperatures and pressures on the surface. Well, perhaps Venus's atmosphere can come to the rescue here as well. At an altitude of around 50 km (30 miles), the temperature and pressure are very similar to those found on the surface of the Earth – and apart from being 50 km up in the air the conditions are far more hospitable than anywhere on the surface of Venus. So if we ever did want to colonise Venus, we might need to take hot air balloons and glide-suits rather than ice packs and thermal insulation. Our ability to build interplanetary airships is a little way off, however, so for now we're stuck gazing at Venus from afar, that wonderful beacon of evening and morning skies.

* 7 *

Earth and Moon

With all this talk of other planets, it can be easy to forget our own planet Earth. In some respects, it's just one of eight planets orbiting a fairly normal star in the outer regions of what seems to be a standard spiral galaxy. But of all the other places we know of in the Universe, this is the only one known to harbour life. Perhaps this is because we haven't looked hard enough, or maybe we just need to be patient. But maybe the Earth is unique in some way, allowing it to play host to millions of species of plants and animals, and enabling the development of advanced, intelligent life.

Orbiting high above is the Moon which, at a quarter the diameter of the Earth, is surprisingly large compared with our relatively modest-sized planet. Though there are larger moons in the Solar System, there are no others which are so large compared with their host planet. Perhaps this makes the Earth–Moon system special in some way. But before we get into speculation about our own self-importance, we should consider what we know (and what we don't) about the Earth and the Moon.

THE EARTH

First, the basics. Earth is a rocky planet that orbits the Sun at a distance of around 150 million km (100 million miles), taking around 365 days (or one year) to complete an orbit. It spins on its axis once every 24 hours, giving rise to night and day. The axis around which it rotates is tilted by 23 degrees relative to its orbit, and so at different times of the year the planet is tipped in different directions relative to the Sun. When the North Pole is tipped towards the Sun, the northern hemisphere experiences summer, while the southern hemisphere feels the cold depths of winter. Half a year later the South Pole is tilted towards the Sun and the North Pole away from it, and the seasons are reversed. These seasonal variations are relatively minor in the grand scheme of things, and tend to only cause swings in temperature of a few tens of degrees at most.

The Earth's atmosphere helps maintain temperatures on the surface which allow liquid water to exist over most of its surface. This atmosphere has currents and circulations in it, which are driven by energy from the Sun and the flow of heat between the atmosphere and the ground. The prevalence of water and life has altered the surface in dramatic ways. Water, particularly when combined with wind, is incredibly efficient at eroding rock, leading to the creation of sand and sediments. The life in these sediments alters it still further, creating the black soil we are familiar with in our gardens. In addition, heat from the interior of the planet causes the surface crust to move around on top of the molten mantle. As well as moving the continents around the surface on a timescale of tens of millions of years, this has also caused much of the surface to be 'recycled'. Parts of the surface are pushed deep into the interior, re-melting under the intense pressures and temperatures. In other places, molten rock flows from within and solidifies, creating new areas of the surface.

Combined with the erosion from wind, water and the effects of life, this tectonic activity means that much of the evidence about the history of the Earth has been erased, either buried under new sediments, or destroyed as regions of the surface are melted at the tectonic plate boundaries. We can, however, calculate the time since the different layers of rock were laid down, by measuring how much the various radioactive elements have decayed in the intervening time. As a result we have an excellent geological history of the Earth, and know that it was created around 4.5 billion years ago. It formed out of a disc of gas and dust that was left over from the formation of the Sun, and which was orbiting around it. As the young Sun's light increased in intensity, the lighter gases, such as hydrogen and helium, were blown into the outer Solar System, leaving just the heavier particles of dust. These particles of dust stuck together to form larger grains, which then clumped to form objects the size of pebbles, rocks and boulders. After millions of years of collisions, some of these rocks had stuck together to form a much larger object, and the gravitational pull allowed this proto-planet to start accreting more and more material. Eventually, almost all the rock in the inner Solar System became part of the four rocky planets: Mercury, Venus, Earth and Mars.

The early Earth was a ball of molten rock, being continuously pummelled by debris flying around the Solar System. Deep within the planet, its own gravitational pull kept the rock molten, and allowed the heavier elements to sink towards the centre of the planet. This is why the core of the Earth is very rich in iron, while the outer crust is dominated by lighter elements such as carbon, oxygen and silicon. Gradually the planet cooled and the rock started to solidify, though there was intense volcanic activity all over the surface for a billion years or so. These volcanoes belched gases onto the surface, forming a pretty foul atmosphere rich in carbon dioxide and with small amounts of other gases such as ammonia, methane and water vapour – not too unlike the atmospheres of

Venus and Mars. More gases and light elements were delivered by the impacts of asteroids and comets, adding to the atmosphere. As the planet cooled, the temperature and pressure fell within the range that allows liquid water to exist. The water vapour started to condense out of the atmosphere, forming the Earth's oceans, rivers and valleys. The oceans had an enormous impact on the composition of the atmosphere, as much of the carbon dioxide dissolved into the water. This relative lack of carbon dioxide in the atmosphere is one of the key differences that distinguishes Earth from Venus and Mars.

THE MOON

Of all the objects in the sky, the Moon is the easiest to observe. In ancient mythology, the Moon is given almost as much significance as the Sun, and with good reason. It is brighter than anything else in the night sky, and can even cast shadows on a clear night. Of course the Moon is also sometimes – in fact about half the time – up during the day, though it is harder to see because the Sun is so much brighter.

The Moon is relatively large, with a diameter of around 3,500 km, making it about a quarter the diameter of the Earth. It orbits the Earth once every 27 days at a distance of around 400,000 km (250,000 miles), and is by far our nearest neighbour. For reasons that we'll return to shortly (see 'Tides', p.133), the same side of the Moon always faces the Earth, so we see the same features from month to month. There are a number of distinctive features on its surface which are easily visible to the naked eye, of which by far the most apparent are the dark areas called the *maria* (Latin for 'seas'). Going by such names as *Mare*

The Full Moon, as seen by the Galileo spacecraft as it passed by Earth.

Tranquillitatis, Oceanus Procellarum and *Sinus Iridium*, these were initially thought to be expanses of water. However, they are now known to be made of dark, low-lying volcanic rock that erupted onto the surface between 3 and 4 billion years ago. The brighter areas are the lunar highlands, which reflect slightly more light. The patterns seen in the lunar seas are many and varied. By far the most famous is the 'Man in the Moon', though this can often be rather hard to see. Other people see a rabbit, or hare, leaping from right to left, with its long ears flowing back around the north-western (top-left) limb of the moon and the isolated *Mare Crisium* ('Sea of Crises') representing its bobtail in the north-east (top-right). Needless to say, these patterns are simply human imagination running wild, though it can be good fun to try to think up new ones.

PHASES OF THE MOON

The Moon doesn't always look the same in the sky, and can often be seen during the day. It also moves around, and will be in noticeably different positions on consecutive nights. This is because the Moon is orbiting the Earth, moving relative to the background stars and even the Sun. Although we always see the same face of the Moon, different amounts of that face are illuminated as the satellite moves around its orbit. The varying appearances of the Moon are called the phases of the Moon and follow a well-defined pattern that can be tracked from night to night, and week to week.

Let's start with the Sun, Earth and Moon roughly in a straight line, with the Moon on the same side of the Earth as the Sun – though it's rarely directly in the way of the Sun as its orbit is slightly tilted. The side of the Moon facing the Earth is, at this point, facing away from the Sun, and so is completely unlit. We call this the 'New Moon', and it is essentially completely dark – it certainly can't be seen against the bright glare of the Sun. Over the next few days, the Moon moves around its orbit, and as it does so we start to be able to see a small part of the side of the Moon facing towards the Sun. This sliver forms a crescent, looking like the curve of a capital 'D' in the sky, and grows bigger over the proceeding days. It is still very close to the Sun in the sky, but will be above the horizon for a short while after sunset, creating beautiful and dramatic views.

Over the subsequent days we see more of the illuminated side of the Moon, and that crescent grows larger. Around a week after

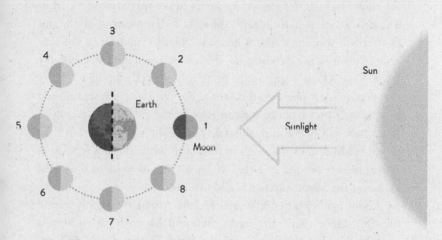

NOTE: Not to scale

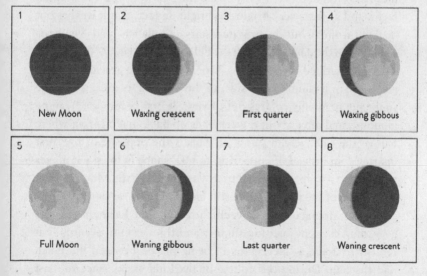

The phases of the Moon as it orbits the Earth.

the New Moon, it has moved a quarter of the way around its orbit. The side we see is half-lit, with the right half appearing bright and the left half seeming dark. When it is exactly a quarter of the way around its orbit the terminator – the line between the night and day sides of the Moon – appears to be almost perfectly straight. This is the 'first quarter' phase of the Moon, though is sometimes referred to as a 'half Moon' since we can see half the side that is facing us (the other half being in darkness). This is the most common time to see the Moon, as it is high in the south when the sun sets. As it moves further around its orbit we see more and more of the sunlit side, and the Moon enters its 'gibbous Moon'.

About a week after first quarter, the Moon is halfway round its orbit. Once again, the Earth, Sun and Moon form an almost straight line, but this time with the Earth in the middle. We are looking at the fully illuminated side of the moon, a phase called the 'Full Moon'. Since it is opposite the Sun in the sky, it is up all night, and appears very bright. So bright, in fact, that it washes out the light from all but the brightest stars and planets, and even casts shadows. In the days following the Full Moon it switches from being what is described as a 'waxing Moon', when we are seeing more and more of its illuminated side, to the 'waning' phases, where the sunlit site gradually rotates out of view. It first becomes a 'waning gibbous', and after about a week ends up back in a position where half of the side we can see is lit. This is the opposite of the 'first quarter', but instead of appearing as a 'D' shape in the sky it makes a filled-in 'C' shape. Referred to as 'last quarter', the Moon rises in the middle of the night, and is still high in the sky at dawn.

As the Moon completes its circuit around the Earth, it becomes a 'waning crescent', getting thinner and thinner until eventually it returns to being a 'New Moon', being placed between the Earth and the Sun. During the course of the four weeks since the last New Moon, the Earth has also moved around its orbit slightly, and the Moon must travel a little bit further around its orbit to 'catch up' with where the Sun is in the sky. This means that although the Moon goes around the Earth in 27 days, there are actually 29 and a half days between two consecutive New Moons. That is almost, but not exactly, the length of a month (which are of course of slightly different lengths anyway), and while there are twelve months in a year there are thirteen Full Moons.

Although we generally only see the parts of the Moon that are illuminated by the Sun, it is sometimes possible to see the night side. When the crescent Moon is seen near the horizon, normally in evening twilight, it is often possible to see the part that is not directly sunlit. Such a sight is sometimes called the 'Old Moon in the New Moon's arms', though is more properly known as 'Earthshine'. In these circumstances, these darker regions of the Moon are in fact shining by the light of the Earth, which reflects a fair amount of sunlight – particularly from the clouds. The appearance of the Earthshine depends on where the Moon is in its orbit (it only happens for a day or two either side of New Moon) and also on how reflective the Earth is, which in turn depends on the cloud cover. There are some beautiful photographs of Earthshine, showing a tiny thin crescent and the dim glow of the Earth-lit parts of the Moon, set against the blue of the twilight sky.

OBSERVING THE MOON

It might seem that the best time to observe the Moon is when it is full. After all, that's when it's at its brightest, and when it's in the sky all night. But that's not necessarily the case, and different features on the Moon are best observed at different phases. The craters, mountains and valleys on the Moon are generally best seen when they are near the terminator since that's when they're illuminated from the side. If you look at the terminator, you'll see it's not a smooth line, but jagged, which tells us straight away that the Moon is not a perfectly round ball. The craters, valleys and hollows just on the daylight side of the terminator are already in shadow, with the Sun having set behind higher features. Conversely, the peaks of mountains which are just on the night side are poking up into the sunlight, creating bright spots. As the terminator moves round the Moon over the course of the 29-day cycle of phases, different parts of the surface become easier to observe. If you're observing the Moon through binoculars or a telescope, an excellent place to start looking is along the terminator. But we can describe some of the features that appear over the course of the lunar cycle.

A few days following New Moon, when it is visible in the evening twilight, the first lunar sea to become visible is *Mare Crisium*. It's a dark, round area, isolated from the other lunar seas, and is seen just above the middle of the eastern (right-most) limb of the Moon. Of course, we have to be careful with directions and orientations on the Moon, as when it's near the horizon it can appear tipped over towards the direction of the Sun. The most reliable way of referring to locations on the Moon is to use north, south, east and west. From the northern hemisphere we see the Moon with its north pole pointing upwards and south downwards. Then, as with the surface of the Earth, east on the Moon is to the right as we look at it, and west is to the left. If it helps, imagine that the Moon is a globe of the Earth, with the UK, Europe and Africa positioned down the centre. In that case, Russia and India (which are in the east) would be over to the right, while North and South America (which are in the west) are to the left.

To the south-west (bottom-left if the Moon is seen 'upright') of *Mare Crisium* are three slightly larger seas which are in an almost straight line, and which blend together at their borders. From south-east to north-west we have *Mare Fecunditatis* ('Sea of Fertility'),

Mare Tranquillitatis ('Sea of Tranquillity'), which is somewhat irregularly shaped, and *Mare Serenitatis* ('Sea of Serenity'). The Sea of Tranquillity is where Neil Armstrong and Buzz Aldrin took mankind's first steps on the Moon in July 1969. The Apollo 11 Lunar Lander is far too small to see, but we can locate the region of the Moon in which they landed. Tranquillity Base, as it is known, is in the south-western part of the Sea of Tranquillity, not far from a very small, bright crater called Moltke (though you will need a small–modest-sized telescope and a keen eye to see this).

If all three of these lunar seas are visible, then look to the north-west edge of *Mare Serenitatis*. A curved line of white spots loops around, and this is one of the largest mountain ranges on the Moon: the Lunar Apennines. These first come into view at around first quarter, with their high peaks casting long shadows and making for one of the most spectacular regions of the Moon to view with a small telescope. The Lunar Apennines lie around the edge of one of the largest lunar seas, *Mare Imbrium* ('Sea of Rain'), which comes into view during the days after first quarter. It has a rather well-defined north-eastern boundary, bordered not only by the Lunar Apennines to the east, but also by the Lunar Alps to the north. The crater Plato can be seen as a dark patch with binoculars, just separated from the northern border of *Mare Imbrium*, which is filled with dark volcanic rocks and is surrounded by high walls that tower 2 km above the base of Plato itself. Further round the north-west of *Mare Imbrium* is a dark, semi-circular protrusion, which is called *Sinus Iridium*. Following the aquatic naming scheme, this translates rather wonderfully and evocatively as 'Bay of Rainbows', and comes into view about three days after first quarter.

There is no clear southern or western boundary to *Mare Imbrium*, as it blends with *Mare Insularum* ('Sea of Islands') to the south and the largest lunar sea, *Oceanus Procellarum* ('Ocean of Storms'), to the west. The first of these, *Mare Insularum*, is perhaps named because it appears to have a large island in the middle of it, though since the lunar seas are in fact nothing to do with water this is clearly not an actual island. Instead, it is an impact crater called Copernicus, and is bright because it was created relatively recently. The crater itself is easier to see when it first appears, a day or two after first quarter, but as the lunar cycle progresses and approaches Full Moon the appearance of Copernicus is dominated by a bright region surrounding it. Look carefully through binoculars or a small telescope and you should be able to pick out lines leading away from Copernicus in all directions. These are called crater rays, and are caused by the material from the original impact being thrown up in the air and away from the crater itself.

Further south there is a much better example of a crater with rays: Tycho. Almost directly on the centre-line of the Moon, the crater comes into view at first quarter. In binoculars, and even to the naked eye, it can be seen as a bright patch towards the south pole of the Moon. Around Full Moon, when the crater is seen illuminated almost straight-on from our point of view, it can appear very bright indeed, and that's when binoculars or a small telescope will start to pick out the bright rays leading away from it. The brightest and easiest to see are the rays closest to the crater itself, particularly those that cut into *Mare Nubium* ('Sea of Clouds') to the north-west, but they dominate much of the southern half of the Moon. Some of these rays are incredibly extensive, with one even cutting across the middle of *Mare Serenitatis* to the north-east. The material that formed the rays must have been flung incredibly high, allowing it to travel almost halfway round the Moon before hitting the surface again. Of the recent craters on the Moon, Tycho and Copernicus are two of the largest, at nearly 90 km across, and therefore the most easily observed. There are many others, though, with a

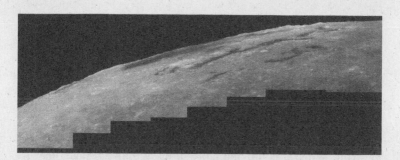

Mare Orientale seen from the ground
by Pete Lawrence.

notable example being Aristarchus, which is the brightest large feature on the Moon – made even more conspicuous because it is located in the middle of the much darker expanse of *Oceanus Procellarum*.

This brings us almost to Full Moon, with one of the last large features to appear being the dark walled plain Grimaldi, to the south-west of *Oceanus Procellarum*. After Full Moon, the opposite side of the terminator starts moving across the Moon's surface. This casts the features into darkness in the same order as they appeared, starting with the eastern features such as *Mare Crisium* and ending with *Oceanus Procellarum* and Grimaldi crater just before New Moon. This does give another chance to see the features illuminated from the other side, though these are somewhat less sociable phases to observe because the Moon is rising later.

If you are a serious telescopic lunar observer, then there are a number of features which are worth looking out for. As well as almost innumerable craters, there are less distinct features such as wrinkle ridges, which are the result of the Moon shrinking slightly as it cooled, and lunar domes, which are the remains of ancient lunar volcanoes. You could also try looking at parts of the far side of the Moon – which is, perhaps surprisingly, possible.

The Moon's orbit is not a perfect circle, but is slightly elliptical, with the distance to the Moon varying by around 10 per cent between minimum and maximum. When it is closer, it moves around its orbit more quickly than on average, while it travels a little more slowly when it is further away. The speed at which it spins around its axis, on the other hand, does not change and so the two rotations are not perfectly in sync for the whole of an orbit. When it is moving more quickly than average, near its closest point to Earth, it is slightly further around its orbit than it would be if it always moved at the same speed. That means that it gets ahead of itself, so to speak, and we are looking at its face at a slight angle, allowing us to see part of its trailing edge. The opposite happens when it is at its furthest from Earth, and its spin is ahead of its orbital position, allowing us to see a small part of its trailing edge. In addition to this effect, the Moon's orbit is tilted with respect to the Earth's, so we are sometimes looking down on it below and sometimes above. The result of these two effects is that the Moon appears to wobble slightly as it orbits the Earth. The effect – known as libration – is small, so we can't see the very far side, but over time it is possible to see around six-tenths of the Moon's surface. Unfortunately, the 'libration areas' are only ever seen at a shallow angle, and so are very hard to map. It was Patrick Moore's accurate maps of these regions that were his greatest contribution to the race to the Moon in the 1960s. He also independently discovered an additional lunar sea right on the western limb of the Moon – though since the official coordinate system of the Moon was flipped by the International Astronomical Union in the 1960s it is called *Mare Orientale* (the 'eastern sea').

Although these features right on the limb of the Moon look very elliptical when viewed from Earth, this is simply because we are observing them at a very shallow angle. In fact, almost every crater and impact basin on the Moon is circular. The same is true of pretty much all craters in the Solar System, and is related to how they are formed. Craters are not formed by the depression made by an incoming object, but rather by the explosion of material that is thrown up by the huge amount of energy released in the impact. The impactor typically comes in at tens of thousands of miles per hour, both compressing the material it hits and causing huge shockwaves. Following the impact, the compressed material is then released, exploding violently and causing a circular crater. The mass and speed of the incoming object defines the size of the crater as well as other aspects of its structure. For example, larger craters often have a central peak and an outer rim, formed from the ejected material which has settled back down again. In the largest craters, these features are replaced with a series of concentric rings, often termed impact basins rather than craters.

TIDES

The reason we always see the same side of the Moon is not, as is sometimes said, that it is not spinning at all. Rather, it spins on its axis once every 27 days – i.e. the same rate at which it orbits the Earth. This is not a coincidence; it is caused by the fact that the Earth and Moon exert gravitational effects on each other, generating an effect we call tides. We normally think of them as tides in the ocean, created by the pull of the Moon, but the same effect also takes place in the ground itself. Similarly, the pull of the Earth causes tides on the Moon and, since the Moon is the smaller of the two bodies, the effect is much stronger. There are also tides caused by the Sun, but the effect is much weaker because that is much further away.

Let's consider the effect on the Moon first of all. Tides are caused because the gravitational pull of an object gets weaker as you move away. This means that, at any one time, the Earth is pulling slightly harder on the closer bits of the Moon than the parts that are further away. It's a fairly small effect, since the more distant parts are only *slightly* further away (by about 1 per cent) than the closer parts, but there is an effect nonetheless. The Moon would initially have been spinning much more rapidly, and so the part that is closest to the Earth would constantly be changing. As a result, the rock of the Moon was constantly being pulled around it rotated by the tidal forces, and since solid rock is rather hard to pull around this caused friction. The action of this tidal friction caused the Moon's rotation to slow down until eventually the same part of the Moon stayed closest to the Earth, which means that the same side has to constantly face us. The same thing has happened with all the moons in the Solar System, which are all 'tidally locked' to the planets that they orbit.

But what about the Earth – has that slowed down as well? In fact it has – and during the time of the dinosaurs the length of the day was one or two hours shorter. This effect can actually be measured here on Earth, by looking at fossils of corals from hundreds of millions of years in the past. The corals grow during the day, but stop at night, and so examining them under a microscope shows a series of fine layers, with each one marking a day. Since they grow more during summer than winter, the spacing of these layers is larger in the summer than the winter, effectively marking out the length of a year. Corals that are growing today have 365 layers in each yearly cycle, since there are 365 days per year. Corals that were forming during the Devonian period, around 400 million years ago, show 400 layers in each yearly cycle, indicating that there were 400 days per year back then. There is no way that the length of a year could have changed over this time period, and so the only thing that can have changed is the length of the day. The only conclusion is therefore that 400 million years ago a day was just 22 hours long, and that at earlier times it was shorter still, with the Earth spinning much faster.

One of the facts about tides which can be confusing is that there are two tides per day, not one. If the high tide in the ocean were thought of as a 'bulge' of water being pulled slightly in the direction of the Moon, then surely it should only happen once every time the Moon is overhead – i.e. once per day. Well, as is so common, the situation is a little more complex. Consider the Earth and Moon isolated in space, with both perfectly stationary and neither rotating with respect to one another. Of course, if they weren't moving, then they would fall together due to their mutual gravitational pull, but this is prevented by the fact that they are orbiting around each other – this effectively pins them at a fixed distance from one another, and so in our simplified situation they are stationary. Next, think of the Earth as being a spherical ball of rock covered in a thin sphere of water. Compared with the main body of the Earth, the water on the Moon-ward side feels a slightly stronger pull, while that on the opposite side feels a slightly weaker one. This change in gravitational pull from one side of the Earth to-the other is called the tidal force, and means that the oceans feel a slight pull away from the Earth's surface on the side facing the Moon and also the one opposite to it. Of course, this doesn't mean that there's a gap between the water and ocean floor, but just that the water 'bunches up' at these points a little more.

We've made a few big assumptions thus far, a major one being that the Earth isn't rotating with respect to the Moon. If we allow the planet to rotate, then each point on the surface passes 'under' the Moon once roughly every 24 hours and 50 minutes – with the additional 50 minutes being added because in one day

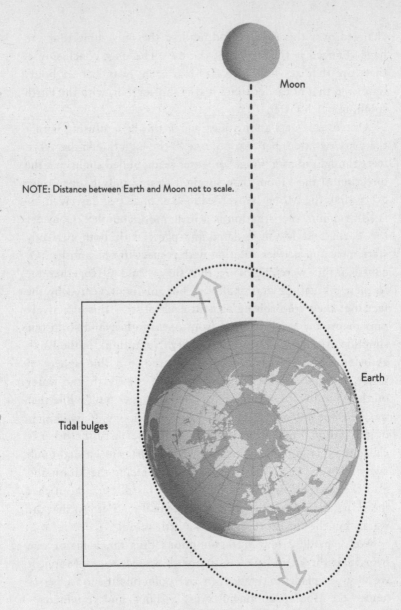

NOTE: Distance between Earth and Moon not to scale.

Moon

Earth

Tidal bulges

The tidal bulges on the Earth due to the Moon.

the Moon moves 1/27th round its orbit, so the Earth must spin a little bit further to line up the same point on its surface with the Moon. The fact that there's also a tidal force on the other side of the planet means that each point on the surface feels a pull once every 12 and a half hours. This tidal bulge sets up a very long, slow wave in the oceans, with the surface moving up and down every 12 and a half hours, moving through the ocean as the planet spins relative to the Moon.

In our simplified version of the Earth, with the oceans forming a thin spherical shell, one might expect this tidal bulge to point directly at the Moon, with high tide occurring either when the Moon is directly overhead, or when it is on the opposite side of the Earth. But if you go to a beach and look up at high tide, the Moon will not be overhead – in fact it's more likely to be near the horizon. This delay is due to the way the ocean behaves when forced to move up and down every 12 and a half hours. The ocean has lots of waves in it, with many different periods and a range of distances between peaks and troughs, but there are certain wave periods at which a simplified ocean would naturally oscillate. These periods are somewhat slower than the tidal period, and so the water is being forced to oscillate faster than it naturally would. The effect of this 'forced oscillation' being faster than the natural frequency, combined with friction between the ocean and the sea floor, means that the tides still move up and down every 12 and a half hours but are out of phase with the Moon, with the tidal bulges directed towards a direction around 30 degrees 'ahead' of the Moon.

This isn't true over the whole planet, because the ocean is quite clearly not a uniform shell all over the surface. We have continents, islands and other land masses, which affect the movement of the tides around the planet. This changes the behaviour of the tides, delaying them in some places and causing them to occur earlier in others. Between Southampton and the Isle of Wight, for example, there are two high tides per day because of the geometry of the English Channel. Typically the tides move the height of the ocean by 1–3 metres, but the topography of some regions means that in some places they are much higher. The highest tides in the world are experienced in the Bay of Fundy in south-eastern Canada, where the sea can move up and down by over 15 metres. Much closer to home, the tidal range in the Severn Estuary can be as high as 15 metres, and the geometry of the estuary means that a single wave, or tidal bore, travelling up the River Severn, can be over 2 metres tall and travels at around 10 miles per hour, with people regularly gathering to surf the bore as it travels upstream.

There are other effects which alter the tide height, the main additional factor being the Sun. Although we are further from the Sun, and so the solar tides are smaller, they do exist. Since the Earth's orbit around the Sun is much slower than the Moon's orbit around the Earth, the solar tides oscillate within a period of 12 hours, rather than 12 and a half, but twice per month they pull in the same direction as the lunar tides. These are the 'spring tides', when the tidal range is highest, and in between them are the 'neap tides', when the pull of the solar tides is perpendicular to that of the lunar tides, and so partially cancels them out.

There are not just tides in the oceans, but also in the atmosphere and even the bulk of the Earth. The pull of all this material causes friction, particularly between the oceans and the sea floor. This tidal friction is what causes the Earth's rotation to slow down, which in turn has a rather surprising side effect. As we've discussed in other chapters, the amount of spin an object, or a collection of objects, has – a quantity called 'angular momentum' – can't easily be changed. In the case of the Earth and Moon we have to consider three spins: the rotation of the Moon, the rotation of the Earth, and the orbit of the Moon around the Earth. (In fact, this last one is a slight simplification since both the Earth and the Moon orbit around their common centre of mass, which lies around 2,000 km below the Earth's surface, but for the purposes of this we can think of the Moon orbiting the Earth.) The total amount of spin of the Earth–Moon system can't change, and so if one of them decreases, another must increase. Over the past few billion years, the Earth has been spinning more and more slowly, with angular momentum being transferred from its rotation to the orbital angular momentum of the Moon around the Earth. Perhaps counter-intuitively, this means that the Moon moves further away from the Earth, and it is currently receding at a rate of just under 4cm per year. Although this is tiny compared with the distance to the Moon, it has been measured very accurately using lasers reflected off mirrors left by the Apollo astronauts, as well as through radar measurements.

All other things being equal, the Moon would continue to slowly spiral away from the Earth and the Earth would continue to slow down. In tens of billions of years the Earth and Moon would be tidally locked to each other, with both rotating at the same rate at which they orbit one another, once every 47 days (where by 'day' we mean the current length of a day). However, in 1–2 billion years the Sun will have increased in brightness and caused the Earth's oceans to evaporate. This will reduce much of the tidal friction that causes the Earth's spin to stop, and the tidal evolution will all but cease. As a result, the Moon will probably end up about 10 per cent further from the Earth than it is now.

ECLIPSES

The Moon's orbit is tilted relative to the Earth's by around 5 degrees. While that's not much, it means that most of the time when it's at New Moon it passes either slightly above or slightly below the Earth. Every now and again, however, the orientation of the

Moon's orbit means that it is in just the right place to move right in front of the Sun. We call such an event a solar eclipse, and they come in a range of types. By far the most famous and spectacular are the total solar eclipses, where the Moon blocks out the entirety of the Sun's disc. What makes them all the more spectacular is that by cosmic coincidence (and it really is just a coincidence) the Moon appears almost exactly the same size in the sky as the disc of the Sun. But, as we saw in Chapter 4, the Sun has an atmosphere, or corona, that stretches far from the main disc we are used to seeing, and eclipses provide an incredible opportunity to observe it.

Most solar eclipses are not total eclipses, however, for a number of reasons. As well as being inclined slightly relative to the Earth's orbit, the Moon's orbit is also slightly elliptical, or elongated. The distance to the Moon ranges between 360,000 km when it is closest to the Earth (called *perigee*) and 400,000 km when it is at its furthest (called *apogee*). Since it is 10 per cent further away at apogee than perigee, it also looks 10 per cent smaller in the sky. If a solar eclipse occurs when the Moon is at apogee then it does not block out the entirety of the Sun's disc, leaving a ring, or annulus. This gives such an eclipse the name 'annular eclipse'. At other times, the alignment is not quite right, and the Moon only cuts off a part of the Sun, forming a partial eclipse.

The eclipse effect is caused by being in the shadow of the Moon, which can be considered in three dimensions. The central part of the shadow, within which the Moon blocks the whole of the Sun, is called the *umbra*, and forms a cone stretching for around 380,000 km behind the Moon. If you are located within the umbra then you will see a total eclipse, but on the Earth's surface this umbra is only a hundred miles or so across, and so you have to be in just the right place. An annular eclipse occurs when the Moon is far enough away from the Earth's surface that the umbra doesn't quite reach it and observers are instead in the *antumbra*, the part directly behind the umbra. Around the umbra is the *penumbra*,

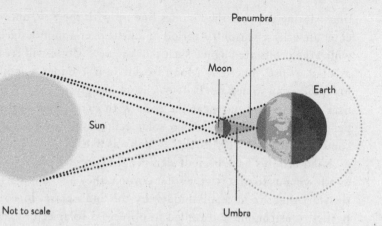

Geometry of a solar eclipse, showing the umbra and penumbra.

within which the Moon only blocks out part of the Sun, and this can stretch for thousands of miles over the Earth's surface. In the majority of eclipses the umbra misses the Earth entirely, passing over either the north or south poles, and so partial eclipses are by far the most common.

To see a total solar eclipse you clearly need to be in just the right place, but you must also be there at just the right time. As the Moon moves around its orbit, the shadow swings across the Earth's surface, with each location typically only seeing totality for a few minutes. For this reason, total eclipses are a very rare sight – unless one happens to be an 'eclipse chaser', traversing the globe to see as many eclipses as possible. But solar eclipses themselves, if one includes annular and partial eclipses, are not as uncommon as might appear. They take place when the Moon crosses the plane of the Earth's orbit, called the ecliptic plane, at or around New Moon. These ecliptic plane crossings are called the 'nodes' of the orbit, and the Moon passes through them twice per month (once going up and once going down). At two times during the year these nodes line up with the Sun, and we enter an 'eclipse season', a period of a little over a month during which the orbits are aligned such that the new Moon will create a solar eclipse. Since the Moon's orbit is just under a month long, there can sometimes be two eclipses in a given season – one near the start and one near the end. With two eclipse seasons per year, there are at least two solar eclipses in any year, and sometimes (though very rarely) as many as five.

Solar eclipses have been of significant historical use. As well as allowing us to observe the solar corona, they also allow us to see stars that are normally hidden by the Sun's glare. In 1919, parties of astronomers travelled to observe a solar eclipse, and measured how the stars appeared to move compared with where they were without the Sun there. The reason for making this rather unusual measurement was to test one of the predictions of Einstein's theory of general relativity, namely that a large mass should cause light to be bent, shifting the apparent positions of background objects. The Sun is indeed pretty massive, though the effect is very small, and so the measurements were hard to make with the portable equipment required. The excursions were largely a success, however, and were one of the first proofs of Einstein's theory.

The nodes of the Moon's orbit are opposite each other, and so if the Moon crosses one of them at New Moon, it will cross the other one at Full Moon. Just as the Earth (or part of it) can pass through the Moon's shadow to create a solar eclipse, the Moon can also pass through the Earth's to create a lunar eclipse. These are just as frequent as solar eclipses, but last longer as the Earth's shadow is four times the size of the Moon's. Since a lunar eclipse is an effect visible on the Moon, anyone who is able see the Moon can observe it, and since a lunar eclipse can last for hours that includes more than half the Earth's surface. Lunar eclipses also come in total and partial varieties, depending on how much

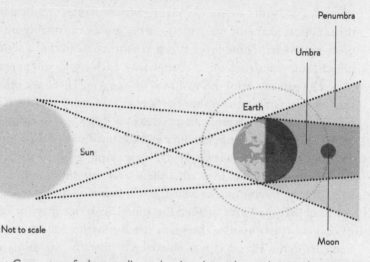

Geometry of a lunar eclipse, showing the umbra and penumbra.

of the Moon passes through the Earth's shadow, but have one additional major difference. The Earth's shadow is not completely dark, with sunlight being refracted by the Earth's atmosphere. Since the blue light is scattered away more easily by the air, what is left filling in the shadow is red light, which is why total lunar eclipses appear red in colour. The darkness of the red colour is an indication of the clarity of the Earth's atmosphere, with deeper colours being associated with more dust in the air, scattering and absorbing more light. Lunar eclipses are of less scientific value these days than solar eclipses, but since they are so much easier to observe they are well worth the wait.

The eclipses do not occur regularly each year for a number of reasons, the first being that there are not a whole number of lunar orbits in a year and so New and Full Moons don't repeat on a yearly cycle. In addition, the tilt of the Moon's orbit rotates, or precesses, due to the influence of the Sun's gravity, causing the nodes to slowly move round the orbit over the course of 18 years or so. To make matters more complicated, the elliptical shape of the Moon's orbit precesses roughly once every nine years. These three effects combined mean that a series of eclipses, occurring on the same day each year, will repeat after 18 years, a period known as a *Saros cycle*. There are other smaller effects that mean that the Saros cycle is not perfect, but we can predict this complicated cycles of eclipses very accurately using our knowledge of geometry and the positions of the Moon and Sun in the sky. Astonishingly, it turns out that the ancient Greeks were also able to predict them over two thousand years ago, despite having no knowledge of geometry, and not even realising that the Earth orbited the Sun. The device they (or one of them at least) used is known as the Antikythera mechanism, named for the island it was discovered near (in the midst of a shipwreck) in around 1900, and was built in the first or second century BC. With a sequence of around thirty cogs, the Antikythera mechanism reproduces the Saros cycle, allowing the user to predict when the next eclipse would begin. At a time when eclipses were associated with dragons eating the Sun and so forth, this represented a great leap forward.

Unfortunately, Europe is passing through a bit of a dry spell in terms of total solar eclipses, and the next one to be visible from the continent will be on 20 March 2015, though even then only from the Faroe Islands to the far north of Scotland. After that, Europeans have to wait until 2 August 2027, though even then only from the southern tip of Spain and Portugal, and if you want to see one from Britain then you need to wait until 23 September 2090. So for the foreseeable future, the only way to see total eclipses will be to travel to far-flung parts of the world – though it does sound like an excellent excuse for a holiday.

THE FORMATION OF THE MOON

It is hard to imagine the Moon not being in the sky, but there could have been a time when it was not there. What is apparent from measurements of its surface is that the Moon was formed very early in the Solar System, and shortly after the Earth itself had formed. There are three general ideas proposed for how the Moon came to be orbiting the Earth. Firstly, it could have formed separately from the Earth and been captured gravitationally at some point in the Solar System's history. The problem with this idea is that it is hard to envisage the Moon approaching the Earth slowly enough to be captured into orbit, as it would be far, far more likely for a passing object to go shooting straight past. A second idea was that the Earth and Moon could have formed together from the same collection of material. If this were true then the two objects should have very similar compositions, but studies of the Moon's interior show that it has a very small iron core. This indicates that overall the Moon has a much lower iron content than the Earth, and therefore is unlikely to have formed in a similar way.

The most popular theory is that the Moon formed after a massive collision very early in the Earth's history, probably within the first hundred million years after the birth of our planet. The impact of an object about the size of Mars (which is roughly half the size of Earth) would have caused a large amount of material to be flung off the Earth and sent into orbit. This material would have coalesced and accreted to form the Moon we see today. This is supported by the fact that the Moon is composed of material more like the Earth's crust than its bulk composition, being relatively iron-poor. There is some disparity in opinions of whether the Moon was formed primarily from material originating on the Earth or the impacting body, but in general this theory is by far the most plausible of those put forward.

The impact of such a large body would certainly have had a huge impact on the early Earth, but there is no way of seeing any evidence on Earth. The effects of wind, rain, tides and geological activity have erased all evidence of such an impact crater. Although the Earth has not experienced such a large impact since the formation of the Moon, we know that massive impacts were much more common in the early Solar System than they are today. For example, many of the large impact craters on the Moon and inner planets seem to date from a period between 3.5 and 4 billion years ago. The basis for this dating uses the principle of superposition, whereby a crater that overlaps another must have been formed later. The same also works for the impact basins in which the lunar maria lie, so we know for example that *Mare Imbrium* is one of the youngest maria, while *Mare Serenitatis* is one of the oldest. To get an absolute date for the impact requires detailed analysis of a rock sample, such as those brought back by the Apollo astronauts. The initial results suggested that these two impacts, and several dozen created in between, were formed within a period around 3.8 billion years ago, but which lasted just a few tens of millions of years. Such a staggering rate of impacts is very surprising this late on in the Solar System's history, since it is hundreds of millions of years after the planets first formed, and raises the possibility of a cataclysmic bombardment of the Moon and inner planets by debris left over from their formation. This period is termed the 'Late Heavy Bombardment', but there are a couple of caveats to be aware of. The first is that the radioactive dating of rock samples only tells us when they most recently solidified, so any impact violent enough to re-melt the rock would 'reset' their ages as calculated through this method. The second caveat is that it is hard to be clear exactly which impact crater the samples originated from, as the larger events scattered material over large fractions of the Moon's surface, and the collected rock may be part of this more recent *ejecta* rather than a previous impact. There is very little doubt that the early Solar System was a very violent place, but there is some debate over whether there was a particularly intense period.

THE FAR SIDE OF THE MOON

The Moon has been tidally locked to the Earth for billions of years, and so until recently no one on Earth had ever seen the far side of it. That all changed in the early years of the space age, when the Soviets and Americans sent spacecraft to take images of the Moon. The first pictures came back from the Soviets' Lunik 3 probe in 1959, and were rather grainy, but technology improved and eventually, in 1968, the astronauts of Apollo 8 were the first humans to ever see the far side of the Moon with their own eyes. What was surprising in those early images was that the far side of the Moon looks remarkably different from the near side. It's still a grey colour, but shows almost no lunar maria. The reasons for this are related to the formation of the maria we see on the near side, and the internal structure of the Moon. The lunar seas were formed following some of the largest impacts to hit the Moon early in its history, which allowed magma from the Moon's molten interior to flood onto the surface. Most of them are known to lie in the centre of massive impact basins, which are present all over the Moon – on both the near side and the far side. In fact, the far side is home to one of the largest impact basins in the Solar System: the South Pole Aitken Basin. The impacts on the far side did not result in such extensive flooding of lava because the crust on the far side is thicker.

The origin of this thick crust is not known, but there are a few theories. One possibility is that tidal forces pulled the molten interior of the Moon towards the Earth, meaning that the mantle is nearer to the surface on the near side. Another possibility is that there used to be a second, smaller satellite of Earth on a similar to orbit to the Moon. Simulations indicate that this would eventually have hit the Moon on the far side, which would have thickened the crust in those regions. The real origin of this lunar dichotomy is not yet known, but may be solved by further analysis of the surface on the far side of the Moon.

MISSIONS TO THE MOON

The main reason for knowing so much about the history of the Moon is that we have been able to travel there and return with samples of rock. Although the Soviets' Luna programme returned a few hundred grams of rock in the 1970s, by far the most useful collection has been brought back by the Apollo missions, totalling almost 400 kg. Although the early Apollo missions were primarily aimed at proving manned lunar missions were possible, the later missions started to include much more science. This is apparent from the choice of landing sites, which were initially based largely on safety and engineering considerations and only later more on scientific grounds. Apollo 11 landed in a relatively flat part of the Sea of Tranquillity, chosen because it was free of large craters and high mountains, and also because if the lunar landing had to be aborted the spacecraft's trajectory would return it straight back to Earth. A similarly flat location was chosen for Apollo 12, this time in *Oceanus Procellarum*, and happened to be very close to where the Surveyor 3 probe had landed two years previously. After the failure of Apollo 13, its landing site was reassigned to Apollo 14, with the intention of sampling material excavated by the impact that created *Mare Imbrium*. This did not mean landing in *Mare Imbrium* itself, but 500 kilometres to the south in the Fra Mauro hills.

The final three Apollo missions were targeted at more scientifically interesting regions, with the landing sites being in slightly more 'adventurous' locations. They also carried a Lunar Rover, which allowed them to travel tens of kilometres across the surface. The first site, for Apollo 15, was a gorge called Hadley Rille, on the boundary between *Mare Imbrium* and the Lunar Appenines. Apollo 16, meanwhile, landed well away from any of the maria in order to collect samples from the Descartes highlands, which represented a type of geology not previously visited. The last manned mission to the Moon, Apollo 17, visited the Tarus-Litttrow valley, on the eastern margins of *Mare Serenitatis*. This area included material much older than that seen before, most of which was related to the creation of *Mare Imbrium*, but also examples of relatively young (well, less than 3 billion years old) volcanism. The mission included the first (and, so far, last) scientist to set foot on the Moon, the geologist Harrison Schmidt.

Manned exploration of the Moon has provided a huge insight into its past, and there have also been significant contributions from unmanned probes. Since 1990 there have been 12 missions to the Moon, launched by NASA, ESA, China, Japan and India. These missions have studied the Moon from orbit, with increasingly high-resolution cameras. One of NASA's most recent missions, the Lunar Reconnaissance Orbiter, has even been able to take images of the Apollo landing sites, showing the lunar modules, the rovers and even the American flags left by the astronauts. Other missions have studied the internal structure of the Moon, showing that its crust is much thinner than previously thought and shedding light on its impact history. As well as studying the geological history and internal structure of the Moon, some of these missions have been searching for water. While water on the Moon would have been boiled off by the Sun's heat, it could exist in the form of ice either on the floors of craters at the poles or under the surface. In 2009 India's Chandrayaan 1 probe discovered the signature of water molecules within the Moon rock, and in the same year NASA's LCROSS mission showed that water does in fact exist in the polar craters after crashing a rocket stage into the Moon and studying the material thrown up.

After the success of the Apollo missions, many people have expressed disappointment that we have not returned, not least some of the Apollo astronauts themselves. One of the reasons for this is cost, with the Apollo missions requiring an unsustainable level of funding from the US government. Although NASA's focus has moved away from the Moon, other nations – particularly China – have expressed an interest in setting up manned bases on the Moon. Likely locations are near the north or south pole, since in some senses these provide the best of both worlds. Some of the mountain-tops near these points are in constant sunlight, and are therefore useful for power generation using solar arrays, while the deepest crater floors likely harbour water ice, which is useful for drinking and making fuel. Living and working on the Moon would certainly be hard, and there is unlikely to be a permanent base there for several decades, but perhaps our children or grandchildren will see it in their lifetimes.

* 8 *

Mars

Once every two years, a conspicuous, baleful visitor takes up a prominent residence in our night skies. This is Mars, the planet that has captured the imagination of human beings for thousands of years. Its distinctive red hue reminded the early Romans of blood, and they associated it with the god of war. The motion of Mars in the sky forced Johannes Kepler to try elliptical orbits (see Chapter 3), and the Victorian astronomers peopled Mars with an ancient but brilliant civilisation trying to save their dying world. Today Mars holds the promise of being the next foothold in the Universe for human beings.

A quick review of the surface hints at why Mars is so fascinating. The inner planets of the Solar System are scorched in the blinding light of the Sun, the gas giants are cold alien worlds beyond the realms of human beings, while the ice giants at the end of the Solar System are even more inhospitable, locked into a perpetual frozen night. Mars is the only place, after the Earth, that holds any promise of being a point of refuge.

A cursory glance at the planet through a telescope suggests a world of deserts and clouds. Recent robotic missions have revealed that many of our early musings on Mars were right. Mars is a world of vast canyons and towering volcanoes. A world of ochre deserts and vast river channels. There seems to be no doubt now that, billions of years ago, Mars was much warmer and wetter than it is now. One cannot help but wonder if any primitive life forms swam in the warm Martian seas, now turned to dust.

Finding the answers to these questions is all part of the ongoing Mars missions, with robotic emissaries combing the Martian soil for signs of life. Amateur astronomers also continue to track the subtle changes in surface features which result from the seasonal dust storms. Mars is slowly revealing its mysteries, essential groundwork perhaps for the first manned missions which are likely to follow in the near future.

ANCIENT OBSERVERS

The brilliant red glow of Mars makes it unmistakable, and no doubt the planet was seen by our prehistoric ancestors. The Babylonians recorded the planet in their observations, and the ancient Egyptian astronomers recorded it and had noticed its retrograde motion in their skies. The ancient Chinese astronomers of the Zhou and Quin dynasties also recorded the movement of Mars in their skies, and a number of conjunctions between Mars and Venus. By the time of the Tang Dynasty they had established the period of Mars's orbit.

Mars's colour meant that the planet was often associated with blood and war – the Babylonians named the planet Nergal, after their god of war and pestilence; similarly, the Greeks called the planet Ares, also a god of war.

TELESCOPIC OBSERVERS

The small apparent size of Mars combined with the chromatic problems of early telescopes meant that the telescopic exploration of Mars revealed very little at first. Galileo was the first person to examine Mars with a telescope, in 1610, but presumably his small telescope revealed nothing about the planet. The Dutch astronomer Christiaan Huygens was the first person to make a realistic recorded observation of Mars.

At around 7 p.m. on 28 November 1659, Huygens turned his rather cumbersome aerial telescope towards the red planet. He was able to make out the dark V-shaped Syrtis Major feature, and this seems to be the feature shown in his drawing from that night. The drawing renders Syrtis Major rather larger than it really is, but given the type of telescope he was using, the result is a good one.

As optics improved, so more astronomers began to train their telescopes on the red planet when it made its appearance in the skies. One of the early facts to be determined about Mars was its rotational period. It was assumed (correctly) that the dark markings seen in the telescope were fixed. So timing how long it took for a feature to return to its first observed position gave an estimate of the rotational period of the planet. Huygens believed the rotational period of Mars to be similar to the Earth's, at around 24 hours long.

Giovanni Cassini observed Mars in 1666 and he determined a rotational period of around 24 hours 40 minutes, which is remarkably close to the real value. The great astronomer Sir William Herschel made a number of observations of the planet between 1777 and 1783, and not only did he derive a reasonably accurate rotational period, but he also estimated the axial tilt of the planet to be 28 degrees. Observing Mars as it passed close to a star, he concluded that if Mars had an atmosphere, it was not very thick.

By the mid 1800s, the first maps of the planet were being compiled. Wilhelm Beer, a German banker and astronomer, working with Johann Mädler, was the first to produce a map of Mars in 1840 using Beer's private observatory in Berlin. It was generally assumed the dark areas of Mars were bodies of water, while the lighter areas were higher land. This is reflected in the wonderful names used on the maps like the 'Hour Glass Sea' (now Syrtis Major), and the Mädler continent. Not everyone thought the dark markings were water; others believed them to be large tracks of vegetation. Giovanni Schiaparelli pointed out large bodies of water should reflect the Sun – a phenomenon that had not been observed.

As we move into the late 1800s, we see the start of one of the most controversial periods in the history of astronomy. In 1877, the Italian astronomer Schiaparelli was using the 8.7-inch refractor at the Brera Observatory in Milan to compile a map (whose nomenclature we still follow today) of the planet. He had observed what he thought were unusual long streaks and straight lines on the Martian surface, and used the Italian world *canali* to describe them. *Canali* is Italian for 'channel', but it was translated as 'canal'. (A channel occurs naturally, canals are manufactured.)

The American Percival Lowell was a wealthy man from Boston; his family was rich and powerful. He studied mathematics at Harvard and travelled extensively in Korea and Japan. He served as a US diplomat in Korea. Lowell read *La Planète Mars* by astronomer and popular science writer Camille Flammarion. This reviewed all the recent Mars observations, including Schiaparelli's. This probably galvanised Lowell into action, and by 1894 he had founded his observatory in Flagstaff Arizona. While other astronomers were uncertain about the canals, Lowell embraced them. He believed that the canals were products of a dying race of Martians, part of a final effort to avoid extinction. He believed the Martians, with their advanced technology, had constructed the vast canal network to transport water from the polar ice caps to the dying cities in the bright ochre deserts. Lowell really *believed*.

When Mars was well placed, Lowell spent as many nights as he could at the eyepiece of his telescope. His notebook contains many hundreds of drawings of Mars, all of them criss-crossed with a magnificent canal network. He wasted no time in publicising his theories; his popular books brought to life the idea that neighbouring Mars was harbouring a dying race. He observed the planet to justify his belief that the canals and their alien manufacturers were real. But as 'Mars mania' swept the world, other astronomers began to doubt the reality of the canals.

The renowned planetary observer Eugène M. Antoniadi (see Chapter 5) provided the real explanation for the canals. During a particularly good Mars opposition, Antoniadi used the 33-inch refractor at Meudon. He watched in amazement as the Earth's atmosphere steadied and he was afforded a magnificent undisturbed view of Mars across the gulf of space. The straight lines dissolved into a bewildering array of fine details and complex structures, all of which looked completely natural, and not at all artificial. Antoniadi realised the canals were simply an illusion, an attempt by the eye to connect a series of fine points. If there was a motivation to see canals, one would undoubtedly see canals.

It took a long time before the canal theory was abandoned. Even in the 1950s, a number of people still believed that the canals might have 'some basis in fact', even if they were natural. The idea that we are not alone in the Universe is a powerful one, and the idea of a dying race building a vast canal network is a wonderful theory – but a very human one. The canal episode is a powerful reminder that it is easy to be fooled by ourselves, and that our personal beliefs and desires have no place in the quest for science.

MARS IN THE SKY

As we have seen, it was the unusual motion of Mars in the night sky that provided Kepler with the inspiration for the elliptical orbit. The elliptical nature of Mars's orbit has a direct impact on how often the planet can be observed in the night sky. The outer planets are best viewed when they come to opposition. If we take Mars as an example, at the time of opposition, Mars, the Earth and the Sun all lie on a straight line and, from our point of view on Earth, Mars appears in the opposite part of the sky to the Sun. The planet will be visible all night and is due south at midnight. (This is true for all of the outer planets.)

Jupiter, Saturn, Uranus and Neptune come to opposition once a year because they orbit the Sun much more slowly than the Earth but, because Mars's orbit takes roughly twice as long as ours, the red planet comes to opposition once about every two years. Interestingly, not every Martian opposition is equally favourable. Mars experiences two types of opposition, *perihelic* and *aphelic*.

Aphelion is that point in Mars's orbit when it is furthest from the Sun. An aphelic opposition occurs when Mars comes to opposition at or near the point of aphelion; because of the tilt of its orbit, this also happens to be when it appears at its most northerly from the point of view of Earth. For northern hemisphere dwellers, this means that Mars will be high in the sky, but its apparent diameter will be small – no more than 14 arc seconds. (By contrast, the Moon has an apparent diameter of around 30 arc seconds.)

Perihelion is the point in Mars's orbit where it is closest to the Sun, and when Mars comes to opposition at or near this point, we have a perihelic opposition. The advantage here is that the disc size of Mars is large – it may be as much as 25 arc seconds. As viewed from the northern hemisphere, however, the planet will be very low down in the southern skies and views of it will disrupted by poor seeing (see Chapter 5, p.89).

OBSERVING MARS

Mars is smaller than the Earth and so we need to know when the best time to observe it is in order to get the best views of the planet. It is quite possible to see the polar ice caps and the prominent dark markings of Mars with a 3-inch telescope when the planet is close to us. In general, small telescopes (3 to 5 inches) will show a number of surface details for a few weeks either side of opposition; larger telescopes will reveal more subtle features and allow you to follow the planet for longer. If you look at the planet well before or well after opposition, the planet does display a slight phase, though the phase never drops below 85 per cent.

From an amateur astronomer's point of view, the most fascinating thing about Mars is the seasonal weather phenomena which occur depending on the type of opposition. Although the atmosphere of Mars is rather insubstantial compared to our own, it is still more than sufficient to produce dust storms and clouds. Normally dust storms occur at around the time of autumn equinox in the northern hemisphere of Mars. A small dust storm looks rather like a small orange cloud; however, they can grow in size. Indeed Mars produces some of the largest dust storms in the Solar System, and a global storm can cover the whole planet, sending a vast amount of orange soil into the atmosphere. When this happens, the dark markings on the surface are obscured, and even a very powerful telescope may only show a featureless ochre disk.

White clouds are another lovely feature of the Martian atmosphere. It is not uncommon for a bright morning or evening cloud to appear in the deep valleys and canyons. The best time to catch white clouds is when the northern Martian hemisphere passes through springtime. The melting of the northern polar ice cap returns water vapour into the atmosphere, and brilliant white clouds collect in the basins and around the volcanoes. When they catch the sunlight, they can look incredibly brilliant.

The polar ice caps of Mars are also rather dynamic places. When the northern hemisphere passes from winter into spring, the brilliant north polar cap starts to sublimate and retreat. Over a period of months, you will see the northern polar cap reduce dramatically in size. In general the southern polar cap seems to be larger, especially when it is winter in the southern hemisphere. As the southern hemisphere passes into spring, the south cap retreats but not evenly, and so the cap appears to fragment and splinter as springtime progresses.

Mars has two satellites, Deimos and Phobos. They are very small and are believed to be captured asteroids. They were discovered by Asaph Hall on 12 August 1877 at the US Naval observatory. Seeing the satellites requires a large telescope, and, even then, the glare from Mars makes them difficult to detect visually.

THE FOUR FACES OF MARS

Mars has a number of striking surface features, and we shall cover them now in a quick tour of the planet's surface. We can split the Martian surface up into four sections, which are sometimes known as the 'four faces of Mars'.

Starting with face number 1, we have the most prominent feature of the Martian surface called Syrtis Major. We now know that this dark 'V'-shaped marking is neither a vast ocean nor vegetation; it is large dark plateau, its dark hues coming from volcanic basalt. The northern edge of Syrtis Major used to be much more pointed; indeed early drawings of the feature show it coming to a well-defined point. Over the decades, however, the dust storms have blunted the edge of it.

In the north is the lovely Boreosyrtis region, while in the south is the Hellas Basin. The Hellas Basin is an enormous

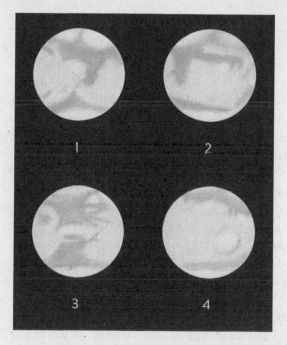

The four faces of Mars. North is at the bottom of each disc,
as it would be seen through a telescope.

impact basin, and there are reasons to believe that it may have
been a small sea when the planet was much warmer and wetter
than it is today. White clouds often collect in the basin and,
when there are a lot of them, Hellas can become dazzlingly
bright. Indeed, on occasions when there is a lot of white cloud
activity in Hellas, the whole of the region can look ablaze as the
clouds reflect the sunshine.

Face number 2 is dominated by the distinctive Sinus Sabaeus
region. This is the long curved streak attached to nearby Syrtis Major.
Sinus Sabaeus ends in Dawes Bay, the line of zero longitude on Mars.
In the north there are a number of bright ochre deserts, including
Eden and Moab while the south contains the Noachis basin.

Face number 3 contains many dark markings. In the north is the dark feature called *Acidalium*, where a large telescope often reveals subtle patches of different intensity and hues. Above *Acidalium* is the bright region called Chryse. Chryse is a basin and it is not uncommon for clouds to collect here making it rather bright. Chryse is currently home to the Viking 1 lander, which touched down on 20 July 1976. In the south is a dark plain, the *Erythraeum* region. To the west of *Erythraeum* is the *Solis Lacus* region, sometimes nicknamed the 'eye of Mars'. In the far south is the vast Argyre impact basin.

Face number 4 has become known as the 'dull face of Mars' thanks to the noticeable lack of any large dark features. There are, however, some interesting features here. Near the equator is the great Tharsis volcano complex. A number of large shield volcanoes can be found here, the largest being *Olympus Mons*, which is three times as tall as Mount Everest. Bright white clouds often collect around the tops of the volcanoes making distinctive 'W' or 'M' shaped clouds. To the west is the bright Elysium plateau. In the north are the only two dark features: *Sirenum* and *Cimmerium*. Above *Cimmerium* are the bright *Electris* and *Eridania* basins, which can also fill with mists and clouds to become quite bright.

It takes about three weeks to see the whole of the Martian surface from any one particular point on the Earth's surface. This is because Mars and Earth have similar lengths of day. On Earth, a day lasts about 24 hours, while on Mars, a day is around 24 hours 39 minutes long. So if you were to see a feature on the central meridian (an imaginary line connecting the north and south poles) at, say, 8 p.m. tonight, those same features will be on the central meridian at 7.21 p.m. the next night, while at 8 p.m. the region of Mars slightly further to the east will be visible. This repeats night after night until, after three weeks, the whole surface has been viewed.

EARLY MISSIONS

Our understanding of Mars increased immensely once we had visited it with space probes. The first successful mission to fly past Mars was NASA's Mariner 4 probe, which flew within 10,000 km of Mars and obtained images of part of the surface with a resolution of 1 kilometre per pixel. These showed no signs of advanced Martian civilisation, as some had hoped, but rather a dry, barren world covered in a very thin atmosphere and peppered with craters. In 1971 Mariner 9 went into Martian orbit, although, in a spectacular case of unfortunate timing, it arrived in the middle of one of the largest global dust storms on Mars, which lasted several months. Eventually the storm abated, and Mariner 9 took images of most of the Martian surface at a resolution of 100 metres. These images revealed the Martian mountains, valleys and craters in unprecedented detail, and tracked weather fronts, clouds and small dust storms moving across the surface. Mariner 9 also revealed one of most fascinating features on the surface of Mars: a vast system of deep canyons that stretches for 4,000 kilometres, or one-fifth of the way around the planet. These canyons are now named *Valles Marineris* ('Mariner Valleys') in honour of Mariner 9. While initially they were thought to be formed by water, in a similar way to the Grand Canyon in Arizona, they are now thought to be a geological feature caused by the stretching and cracking of the surface, similar to the Great Rift Valley that runs down the eastern side of Africa here on Earth. It is thought that this immense tear in the Martian crust formed as the planet cooled and is possibly related to the Tharsis region to the west, which contains some of the largest volcanoes in the Solar System, including Olympus Mons. The Mariner 9 images also showed the presence of what looked like many other ancient river valleys, raising the intriguing possibility that Mars may once have been a warmer, wetter world. Later results have shown that, while

Valles Marineris as seen by the Viking probes.

Valles Marineris was probably formed by tectonic processes, it has been shaped by the effects of water erosion and landslides. The study of the surface and atmosphere, and the possible presence of water – and therefore life – paved the way for what came next: landing on Mars.

In 1971, NASA launched the Viking probes, which would orbit Mars and send landers down to the surface. The missions were a great success, with the Viking orbiters providing maps of the entire surface, showing complex networks of ancient rivers and streams, all of which are now dried up, which could only really have been created by flowing water in Mars's distant past. The flowing of material around some impact craters also suggests that it has impacted mud, rather than dry soil, also hinting at a much wetter environment in the past. The probes also spotted a range of interesting features, perhaps the most famous of which was the 'face on Mars', a collection of rocks in the Cydonia region that, under the right lighting conditions, looks like a human skull. There were countless theories as to its origin but – to the disappointment of some – this was later shown to be just a small pile of rubble.

But the main reason for sending the Viking probes was their landers, which successfully set down on the red planet in 1976. Such an operation is surprisingly difficult due to Mars's thin atmosphere, which is thick enough to cause significant heating during entry, but not thick enough to significantly slow the craft down (something called aerobraking) or allow the efficient use of parachutes. The Viking landers therefore required a combination of aerobraking with a heat shield, parachutes and rockets to gently lower them to the surface. Both landings were a complete success, with Viking 1 landing in the Chryse Planitia and Viking 2 in Utopia Planitia, slightly further north and on the opposite side of the planet. The landers carried experiments which studied the chemical composition of the surface and atmosphere of Mars, but were there for one main reason: to search for life. Each lander carried four experiments targeted at this goal, but the results were inconclusive. One technique used was the baking of samples of Martian soil in order to analyse the gases released, particularly the products of organic chemicals – that is, compounds containing carbon, hydrogen and oxygen, which are thought to be essential for life on Earth. Another type of test was to look for whether carbon introduced to samples of either the soil or atmosphere from Mars was converted into carbon dioxide, which would indicate that something – such as microbial life – was metabolising the carbon. Of the four experiments, three came back negative but one, which looked for the consumption of carbon and release of carbon dioxide, reported a positive result, leading to an overall inconclusive result. It was pointed out by some that this is the same result as that expected from the same experiments conducted in Antarctica, where microbial life exists within the rock but there is no detectable presence of organic chemicals. Decades later, the Phoenix mission discovered a type of salt called perchlorate in the soil near the Martian north pole. This compound is very efficient at oxidising the soil, destroying organic molecules and producing gases such as carbon dioxide. Such an oxidising agent, with the associated lack of organic chemicals, would also mean that microbial life on the surface would be very difficult.

THE CURSE OF MARS

Despite the successes of the NASA missions to Mars in the 1970s, particularly the Mariner and Viking probes, the Soviet programme was far less successful, with over a dozen missions in the 1960s and 1970s ending in failure or only partial success. The most successful was 'Mars 3', which made the first successful landing on Mars, but ceased to transmit after just 15 seconds. The 'curse of Mars', which was light-heartedly attributed to Martians protecting their home world, is a stark reminder that interplanetary travel is a complicated and tricky business.

NASA took a break from Mars in the 1980s, but resumed their efforts in earnest in the 1990s. Of five missions in that decade, only two were successful: the Pathfinder mission, with its tiny tea-tray-sized Sojourner rover, and Mars Global Surveyor, which made global maps of Mars for nearly a decade. The failures of the other three missions were either unlucky or embarrassing. Mars Observer disappeared just before arrival in orbit in 1993, presumably as a result of a failure on the spacecraft. Mars

Climate Orbiter braked too sharply when entering orbit in 1998 and burned up in the Martian atmosphere, the result of a mix-up between metric and imperial units in the calculations for the thrusters firings. Mars Polar Lander, which crashed onto the surface early the next year, was only slightly less embarrassing, with the failure attributed to the jolt of the landing legs deploying triggering the engine shutoff prematurely.

But NASA's orbital programme was much more successful in the first decade of the 21st century. Mars Odyssey, named because of its arrival in 2001 (after the Stanley Kubrick / Arthur C. Clarke film and novel *2001: A Space Odyssey*), is still operational 12 years later (at least, it is at time of writing), and has played a key role not only in investigating the Martian surface, but also in providing a communications relay for missions on the ground. In 2006, Mars Reconnaissance Orbiter, the most advanced Mars orbiter to date, arrived around Mars. Its high-resolution cameras have allowed it to map the surface of Mars down to a resolution of around 30 cm per pixel, not only revealing intricate details of structures on the surface, but also allowing us to see images of other missions in the process of landing on another planet. The staggering photo of the Phoenix lander dangling from its parachute is one of the most remarkable images of the space age.

In 2003, the NASA fleet was joined by Europe's Mars Express, which released the British-built Beagle 2 lander. Unfortunately Beagle 2 was never heard from again, and its fate remains a mystery. Sadly, Beagle 2 was not the last craft to suffer from the curse of Mars, with the Russian Phobos-Grunt mission failing to leave Earth orbit in late 2011. This was reported to be due to cost-cutting. Future missions involve European, Indian and American missions to Mars, launching over the next few years (though by the time you read this some may well be en route, or even have arrived at Mars), with the aim of studying the Martian atmosphere and geology, as well as resuming the search for life.

LOCAL EXPLORERS

The orbiters and landers on Mars have returned a whole host of fascinating results, but by far the greatest publicity has been from the rovers, which have been able to move over the planet's surface. The first was the Sojourner rover, part of the 1997 Pathfinder mission, which landed in a rock-strewn valley 800 km from Viking 1. While the Pathfinder lander itself analysed properties of Mars's atmosphere, Sojourner drove up to a number of rocks over the course of the three-month lifetime of the mission. As well as taking photographs, the rover studied the detailed composition and structure of the rocks, showing that while most were volcanic in origin, some appeared to have been formed as part of a sedimentary process on the bottom of a lake or ocean. In fact, the Pathfinder observations confirmed that the rocks ended up in the valley as a result of a catastrophic flood billions of years ago.

Following the success of Pathfinder and Sojourner, NASA sent two Mars Exploration Rovers to Mars in 2004. Landing on opposite sides of the planet, the rovers Spirit and Opportunity arrived on the surface in the same way as Pathfinder had seven years earlier – by bouncing on giant inflatable airbags. The expectation was that dust would settle on the solar panels, gradually reducing the available power and limiting the missions to a three-month lifespan, and a distance of just 600 metres. But these two plucky rovers far outlived their expectations, thanks to gusts of wind blowing the dust off and extending their lives. Over the course of six years, Spirit drove over 7 kilometres until it got stuck in soft sand in 2009, and finally failed during the Martian winter of 2010. Opportunity, meanwhile, is still going strong (at time of writing), looking to break the record of 37 km set by Russia's Lunokhod 2 rover, which explored the Moon back in 1973. This ability to cover large distances meant that Spirit and Opportunity could pick the best, or at least most interesting, targets, as well as allowing them to travel to regions too hilly or rocky for spacecraft to land in directly. While previous efforts had searched directly for evidence of life on Mars, with the Mars Exploration Rovers the idea was to answer an easier, but just as important, question and go in search of water.

Spirit landed in Gusev crater, which was probably created by an impact over 3 billion years ago and filled or partially filled with water for some of its history. The crater was largely filled with volcanic rock, the result of flooding by lava, but measurements of the rocks and soil by Spirit confirmed that the rocks had certainly been altered by the presence of water, both due to weathering and alteration of the minerals in the rocks. But it was when Spirit climbed the rolling Columbia Hills that some of its most remarkable discoveries were made. The minerals found in some of the rocks here, as well as the presence of sulphur-rich material, indicated that they must have formed in the presence of water, as well as been altered by them over the intervening period. A couple of years into its supposedly 90-day mission, one of Spirit's six wheels stopped working, and for the remainder of its mission it had to drive backwards, dragging the unmoving wheel behind it. Although this played a part in Spirit's eventual fate, making it unable to work itself free from a sand-trap, the dragged wheel helped with one of the rover's most surprising discoveries. In March 2007 the dead wheel ploughed through some soft soil, revealing the bright colours of pure silica beneath. This silica indicates that at some point in Mars's history hot water was flowing on or under the surface, quite possibly in the form of a hot spring. The presence of hot water on early Mars is very promising for the potential for the development of life, though Spirit didn't carry the equipment necessary to search for it.

Opportunity landed in the flat plains of Meridiani Planum, by chance right in the middle of a small 20-metre diameter crater. After spending much of its initial 90-day mission within what was dubbed 'Eagle crater', Opportunity managing to climb out onto the vast Meridiani planes Over the course of the next two years, the rover drove over the plains, dropping into the 130-metre diameter Endurance crater before re-emerging six months later. These craters provide ideal sites for studying Mars's past, as they have excavated the surface and revealed layers of rock laid down earlier on in the planet's history. Following the explorations of Eagle and Endurance craters, mission operators targeted the much larger Victoria crater, 6 km to the south and almost 800 metres in diameter. Opportunity spent two years at Victoria crater, for half of which it was down inside the rim, and established that the geological history was similar to what was seen further north at Endurance and Endeavour. By this point, it was clear that this large region on Mars had once been covered in water, which had shaped and altered the rocks at some point in their past, either as they were deposited or at a later time. Following a rather hairy scramble out of Victoria crater, Opportunity set its sights on the largest crater in the region, the 20-km-wide Endeavour crater. The journey was nearly 20 km long, circumnavigating some small craters and a large dune field, and took three years to complete. The rover, by now far out of warranty, stopped off at anything that might prove interesting en route, including a number of meteorites and a rock that probably came from a different part of Mars as part of the ejecta from another impact. By the time it reached the rim in August 2011, Opportunity had been operating on Mars for six and a half years, far longer than its planned three-month mission. The rocks around the rim of Endeavour were probably placed there after being blown out by the original impact and, since the crater is much deeper than any visited before, were probably laid down much earlier. The initial observations seem to agree, with the material around Endeavour looking similar to other volcanic rocks found by Opportunity, but have a slightly different chemical composition. In particular, they contain veins of gypsum, a mineral that is fairly commonplace on Mars, and which

Victoria Crater as seen by the Opportunity rover.

must be formed in the presence of water. The other rocks found so far could have been altered at a later date, but this gypsum is the first evidence that Opportunity is studying rocks that were laid down in the presence of water.

The discoveries by Spirit and Opportunity have been remarkable, largely thanks to their longevity but also because they carry a sophisticated suite of experiments for investigating the physical properties of Mars. Their successor, however, is a huge leap forwards. Mars Science Laboratory, consisting of the Curiosity rover, is the size of a small car, weighs in at just under a tonne, and carries ten experiments designed to look at the chemistry of the Martian surface. It landed in August 2012 using a novel technique which combined aerobraking in the atmosphere, the use of a parachute, and finally a 'sky crane', which hovered above the surface on thrusters before flying off to crash a few hundred metres away. Using this powered descent allowed Curiosity to be five times heavier than its predecessors, while having the thrusters on the overhead jetpack meant that it landed on its wheels, ready and raring to rove. Rather than being solar powered, Curiosity uses a radioisotope thermal generator, which generates power from the decay of radioactive plutonium. This is similar to the power generation on the Viking landers and any probes that have ventured into the outer Solar System and, while it doesn't generate any more power than Spirit and Opportunity's solar panels, it allows the rover to operate at night and is not susceptible to the build-up of dust. This allows the mission lifetime to be much longer, at two years, with a planned driving distance of 20 kilometres – though if previous missions are anything to go by it may well outperform on both of those criteria.

Curiosity's instruments include ChemCam, a powerful laser capable of vaporising small amounts of Martian dust and rock up to 7 metres away, which is analysed by a detector that measures the composition of the vaporised material. This not only means that Curiosity can study rocks that might be in inaccessible locations (such as up a cliff face), but also allows the mission operators to pre-select the rocks they want to drive up to and study further. At the end of its 2-metre robot arm are a camera capable of taking microscopic image, an X-ray spectrometer that allows the composition of the rocks to be analysed, and a set of tools for drilling into the surface and collecting material. This material is then transferred to the chemical laboratories in the main body of the rover, where advanced instruments analyse the minerals in the samples and look for the presence of organic compounds. These organic chemicals, which contain a combination of carbon, hydrogen and oxygen, are thought to be a prerequisite of life, and Curiosity's chemistry experiments will be able to detect whether the compounds are present, and determine the conditions in which the rocks they are found in were formed. It is this ability which will help Curiosity succeed in its primary mission, which is to determine whether early Mars was a habitable place or not, though it is unlikely to be able to determine whether there was life present – that will have to wait for future missions.

Like Spirit and Opportunity, Curiosity is looking at the layers of rock to determine the geological history of Mars. It landed in Gale crater, an impact crater around 150 km across which has since been filled with sedimentary rock, almost certainly laid down at the bottom of an ancient ocean. The central peak of the crater is Curiosity's eventual goal, and it will climb the slopes to move through the sedimentary layers, building up a geochemical history of the environment on Mars. It has already established that the rocks it studied first were laid down during a phase when Mars

Self-portrait of Curiosity.

was indeed habitable, at least in that location, but investigation of the full geological sequence should allow Curiosity to determine for how much of Mars's history this habitable environment was present. The answers will help drive the experiments of the next generation of Mars rovers, with a European rover due to launch in 2018, and NASA intending to launch a follow-up to Curiosity in 2020.

MARS: PARADISE LOST

What has become increasingly apparent over more than four decades of Mars exploration is that it has not always been the dry, dead planet that it is today. This started with the observations from

Mariner 9 of features that looked very much as if they had been carved by water. A few years later the Viking probes revealed the stark difference between its northern and southern hemispheres. The southern hemisphere is on average a couple of kilometres higher than the northern hemisphere, and is far more heavily cratered. This implies that the southern hemisphere is much older than the northern hemisphere, though the reasons for this are not yet fully understood. One possibility that was initially suggested was a single massive impact which blasted away the crust of the northern hemisphere. If the northern lowlands are an ancient impact basin, it would be the largest example in the Solar System, but there are a number of reasons why that is thought to be unlikely. For a start, the basin is not circular, and also there is no evidence of ejecta – material thrown up from the impact – either around the edge of the basin or over the southern hemisphere.

A more recent theory is that the dichotomy of Mars is due to plate tectonics. It is clear that Mars used to be geologically active, since it has some of the largest volcanoes in the Solar System. There is evidence in the older parts of the surface that ancient Mars had a magnetic field, which would have required convection in the liquid core and therefore some sort of cooling mechanism, which plate tectonics would have provided. The reason the evidence for tectonic activity is not as apparent on Earth is that it must have stopped billions of years ago, and also that it was much simpler. One idea is that there was an upwelling in the mantle in the northern hemisphere which, combined with a corresponding downwelling in the southern hemisphere, resulted in two distinct regions. New, thinner layers of crust would have been created above the upwelling, spreading out and being compressed under the thicker crust to the south. Along the boundary there would have been compression and subduction, which could well be the cause of the Tharsis bulge and the massive volcanoes, including Olympus Mons. Such geological stresses could have also formed the Valles Marineris rift valley. Although this theory has promise, without a much better understanding of the geological history and internal structure of Mars it is hard to prove it one way or the other, though NASA's 'Insight' mission may help provide some answers when it arrives towards the end of this decade to study the planet's interior.

As is common with many bodies in the Solar System, it is possible to work out the order in which various regions of Mars's surface were formed by counting the number of craters – the more cratered a surface is, the longer it has been exposed to impacts and therefore the older it is. Without samples of Martian rocks in labs here on Earth, it is very difficult to establish the absolute date at which rocks on Mars were formed, but by comparing the cratering with that on the Moon it is possible to estimate it to within a few hundred million years. As such, we now have a reasonable idea of the geological history of Mars, and how the planet has changed over the past few billion years. As well as the evidence from the surface, we have over a hundred meteorites known to come from Mars, many of which show evidence of containing water. These can be dated more reliably, and show that the rocks formed in the presence of water around 2 billion years ago. Of course, since these meteorites were found on Earth we have no way of knowing which part of Mars they came from.

There is an incredible amount of evidence for the presence of flowing water on Mars at various points in its past. Much of this evidence comes from images of valleys and river channels on Mars, which look very much as if they were carved by flowing water. Some of these valleys are very recent, just tens of millions of years old, indicating that water has been flowing on Mars in relatively large quantities in the relatively recent past. There are pockets of water ice seen in the floors of some craters and valleys, and significant quantities are locked up in the polar ice caps. Recent observations of various odd structures on the surface are thought to be caused by water ice freezing and evaporating as the seasons change, leaving depressions in the overlying soil. There is strong evidence that water ice is stored underground on the planet, possibly relatively close to the surface, and it is the release of this water which is thought to have led to periods of intermittent flooding on Mars. This periodic release of water could be linked to massive changes in Mars's climate and seasonal variation, possibly linked to the fact that the tilt of Mars's axis has changed over the past few hundreds of millions of years. Mars's axis is tilted at around 25 degrees to the plane of its orbit, not too far from the 23-degree tilt of the Earth's axis. But while Earth's axis has remained stable, probably due to the presence of its large Moon, Mars's has tipped and tilted over time, and there is evidence that over the course of just hundreds of thousands of years the axial tilt of Mars can vary by tens of degrees. Such a massive change in the tilt of the axis would have a strong effect on the climate and seasons, with the polar ice caps not only growing and shrinking but also moving around over the planet. It seems that these wet periods of Mars were intermittent and only lasted for short periods – probably not long enough for life to evolve and develop. But while the recent history of Mars has largely been completely dry, there is also strong evidence that around 3.5 billion years ago Mars was much more like the Earth, with a thicker atmosphere and oceans of water. It may even have been able to harbour life – a question that the Curiosity rover is attempting to answer. The fate of the Martian atmosphere, which is thought to be only 1 per cent of its original thickness, is the target of Maven, a NASA mission due to arrive at Mars in 2014.

Although the water on Mars is mostly locked up in the ice caps and beneath the surface, the pressure on the surface is too low for it to persist in a liquid state. However, observations from the Mars Reconnaissance Orbiter have shown that it may be able to exist for short periods, possibly staying melted by being mixed with Martian soil, much as salt is used to prevent ice on roads on Earth. Dark-coloured gullies around the rims of craters appear and fade with the seasons, with the interpretation being that the Sun has heated the surface and released salty water. This water has briefly dampened and darkened the surface before evaporating after just a few hours. This is the most promising sign that flowing water exists on the Martian surface today, albeit briefly.

Given the similarities between ancient Mars and the Earth, the question remains: was life able to form on the red planet? There is no evidence for advanced life, but it is possible that microbial life formed billions of years ago. Experiments with bacteria in space have shown that such primitive life doesn't necessarily die when the conditions become unfavourable, but might simply hibernate, thriving once again when conditions allow. The possibility therefore exists that this life could still be there today. Perhaps it exists near the underground reservoirs of water ice, which, as we've seen, do melt from time to time. The signature of microbial life, past or present, is very hard to detect – as the Viking missions showed in the 1970s. Most techniques rely on searching for the waste products of life, which on Earth include compounds such as carbon dioxide and methane. The Martian atmosphere is almost entirely composed of carbon dioxide, so small increases due to life are hard to detect, but methane is much less common. Part of the reason for this is that methane should be quickly split up by sunlight and lost from the Martian atmosphere. Recent evidence has shown that there are very small traces of methane in the atmosphere, though how localised these traces are and how long they exist for is yet to be determined.

There are two possible explanations for this methane, with one being some sort of biological life and the other being very weak geological activity. Either one would be exciting, as Mars shows all the signs of being dead in both senses. The Curiosity

Rover is sampling the atmosphere at its location in Gale crater, but to fully understand it a detailed study of the atmosphere is required. As well as the NASA's Maven spacecraft, the European-led ExoMars mission will include the Trace Gas Orbiter which, following its launch in 2016, will investigate the presence and evolution of methane, along with other gases, with an aim of establishing its origin. The intention is that this investigation of the atmosphere will help pick the landing site for the second phase of the mission: a rover. A collaboration between ESA and the Russian space agency, the ExoMars rover will be much smaller than Curiosity, but designed specifically to search for evidence of life, either past or present. So, with any luck, by the end of the decade we'll know the answer to the age-old question: is there life on Mars?

9

Jupiter

Beyond the orbit of Mars, after the rubble heap of the asteroid belt, we find a totally different kind of world: the *gas giant*. Compared to the terrestrial planets, Jupiter and Saturn are very alien worlds. Jupiter is the larger of the two – it is a colossal ball of hydrogen and helium whose extensive gravitational field keeps a firm hold of some 67 satellites.

The first sight of Jupiter through a telescope changed for ever the way we thought the Universe worked, and in doing so ushered in the beginnings of modern astronomical science. As we shall see, telescopic observations of the planet hinted at its unusual characteristics – the colourful clouds of the planet's equator rotated more quickly than the rest of the planet. The realisation that the planet had a thick and extensive atmosphere came early on as the telescopic observers faithfully tracked and recorded the many storms which constantly rage in the Jovian atmosphere.

The early spacecraft to Jupiter, Pioneer and Voyager, revealed facts about the Jovian system never dreamed of – the volcanically active moon Io being perhaps the greatest surprise of all. Jupiter has probably been one of the most surveyed planets in the Solar

System, and yet amateur astronomers continue to play their part: in the early 21st century we have seen a number of comet impacts with Jupiter, all of which were discovered by amateurs. Professional astronomers too have found the spacecraft data have left more questions than answers. Jupiter is going to keep us busy for a long time – quite apt for a world that is never still.

ANCIENT OBSERVATIONS

Jupiter is one of the brightest objects in the night sky, and so there is no way it could have been missed by the ancient peoples. We know that the Babylonians recorded their observations of the planet, but recently David Hughes and then Chinese astronomical historian Xi Zezong have suggested that the satellite Ganymede was recorded by the ancient Chinese astronomer Gan De with the naked eye. Since these observations were recorded in 362 BC, this would predate the discovery of Jupiter's moons by Galileo by quite some margin. In fact, this is not as unrealistic as it first seems; Ganymede is a large satellite and, with an apparent magnitude of around 5.0, it is within naked-eye visibility. The only problem is the glare from Jupiter, yet even today a number of people claim they can see the satellite with the unaided eye.

Without a telescope, Jupiter revealed little else about the Solar System, but when the first telescope was turned on to the planet in 1610, it marked the beginning of a new chapter in astronomical history.

TELESCOPIC OBSERVERS

As we saw in Chapter 3, the first person to examine Jupiter with a telescope was Galileo, an event which had a profound effect on human history. As other astronomers pointed their telescopes towards Jupiter, they began to notice more details. Christiaan Huygens was a Dutch natural philosopher who established a long aerial telescope to examine the heavens. A drawing made by Huygens in 1659 clearly shows two equatorial belts circling the equator.

By 1665, the Italian astronomer Giovanni Cassini had turned his attention towards Jupiter. Cassini and the English natural philosopher Robert Hooke independently observed the Great Red Spot for the first time. Cassini used the Great Red Spot to make an estimate of Jupiter's rotation which he found to be some 9 hours 55 minutes. He also realised that the planet rotated more quickly at its equator than it does at the poles, a phenomenon known as *differential rotation*.

By the 19th century, telescopes were becoming much better. More accurate tools and techniques allowed refractor telescopes to be made with less chromatic aberration and at much shorter focal lengths, while they also became much easier to handle. Newtonian reflectors were becoming more common, too. Many astronomers of the day made their own mirrors – Sir William Herschel was particularly good at mirror-making.

In 1865, the British astronomer Warren De la Rue showed that the cloud features of Jupiter were not identical, with many of them showing different colours. Other astronomers of the late 19th and 20th centuries continued to observe the planet, charting its continually raging storms.

THE FEATURES OF JUPITER

A brief look through a small telescope reveals a planet that looks very different to our own. It is appreciably oblate, and its colourful cloud tops are arranged into a series of dark brownish coloured belts, and bright yellowish zones. Vast Earth-sized storms swirl and rage in the belts and zones and, because Jupiter has such a short day, it doesn't take long for the various features to move across the disc.

It is believed that the dark belts are warmer clouds of hydrogen sulphide (and other things) lower down in the atmosphere, while the brighter zones are thought to be ammonia ice crystals high up in the colder parts of the Jovian atmosphere. A closer look at the belts and zones often reveals the existence of oval storms – vast hurricanes that have been blowing for decades. A number of Jupiter's storms are long-lived; on Earth, hurricanes start out at sea and lose energy when they hit land, but on Jupiter there is no land so once a storm gets started, it can go on for a long time.

One storm in particular which has lasted for hundreds of years (possibly longer) is the Great Red Spot. Through a telescope, the Great Red Spot looks like a salmon-coloured oval; drawings from hundreds of years ago show it to be larger, and redder than it is today. A number of theories were advanced as to what the Spot might be. One favourite was a solid body floating in the atmosphere of Jupiter. Another theory correctly suggested it was a vast hurricane. The debate was settled once and for all when the Voyager spacecraft imaged the feature in detail and showed it whirling and churning in the Jovian atmosphere. The Spot is enormous, and could easily swallow three planet Earths. The Great Red Spot does not seem to be constant in size; rather, it seems to be shrinking and, at its present rate of contraction, it could be circular by 2040 (although Jupiter's powerful winds will probably not allow it to stay circular). We don't really know

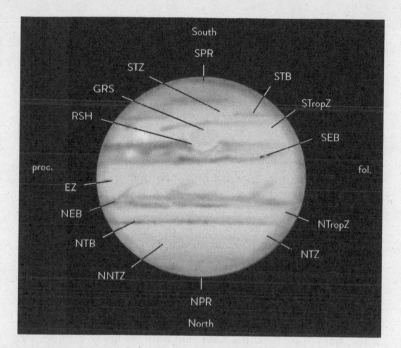

The names of the belts and zones of Jupiter as seen
through a telescope, with north down.

how long the Spot will last for, so if you have a telescope, try to
catch a glimpse of it.

When we look at Jupiter through a telescope, all we are ever seeing
is the top layer of a very extensive atmosphere. As you might expect,
this makes the geography of Jupiter rather challenging. Nonetheless,
astronomers have devised a useful reference system which provides
names for the various belts and zones (see diagram above).

Through a small telescope, the northern polar region of the
planet looks like a mottled greyish-brown region. Larger telescopes
and high-resolution images reveal the presence of faint oval storms.
Below this is the bright North Temperate Zone (NTZ), and just
below this is the North Temperate Belt (NTB). The NTB has recently

shown some interesting changes both in colour and intensity, and it seems to be a rather dynamic feature. Below the NTB is another bright zone, the North Tropical Zone (NTropZ). Normally, this is a yellow-white zone, but in mid 2012 the zone darkened and became a ruddy yellow colour, and this was apparent even through a small telescope. By the end of 2012, the zone had returned to its usual appearance. Quite why the NTropZ did this is still a mystery.

Continuing further southwards, we come to one of the main belts of Jupiter – the Northern Equatorial Belt (NEB). This belt is quite prominent and can be seen through a 3-inch telescope. The edges of this belt often contain a number of bluish streaks called festoons which extend into the Equatorial Zone. The NEB is usually the most active belt on Jupiter – it is not uncommon for dark brownish ovals (called 'barges') to appear in the belt, and larger telescopes and high-resolution images have shown all manner of storms and fine structures constantly evolving in the NEB.

Through a small telescope, the Equatorial Zone (EZ) is a bright white, but larger telescopes reveal not only the bluish festoons but sometimes a faint equatorial belt. At times the EZ can become quite dark and full of storms.

Continuing south we come to the Southern Equatorial Belt (SEB). Like the NEB, this is another prominent dark belt which can usually be seen through a 3-inch telescope. Unlike the NEB, however, the SEB is not always prominent. There are times when the belt fades away completely. After a period of time, a series of white spots erupt where the SEB previously was, and an event called an 'SEB Revival' occurs: dark material erupts from the spots and the material is carried by Jupiter's winds and the belt dramatically reappears.

Below the SEB we come to the South Tropical Zone (STropZ). This is another bright zone, and through a small telescope it appears to be a yellow-white colour. Larger telescopes show that the zone has a number of subtle features, including a faint belt called the South Tropical Band, along with various spots. The South Tropical

Zone is also home to the Great Red Spot, which sits just below the SEB and makes an indentation into the belt called the Red Spot Hollow – this is visible through a telescope of 4 inches or more. Interestingly, when the SEB fades, the Great Red Spot gets darker and redder, and when the SEB returns the Great Red Spot fades again. At the time of writing it is a light pinkish colour and can be picked up with a 4-inch telescope.

Continuing south, we come to the next belt, the South Temperate Belt (STB). Normally a brownish-grey in colour, it can be quite dark, and easier to see than the NEB. The STB is home to a number of long-lived storms including white ovals, and a feature known as 'Red Spot Junior' (or officially as Oval BA) which, as its name suggests, looks like a small red spot. The white ovals and oval BA are hard to pick up visually and require a large telescope to see them.

Finally, we come to the southern polar regions (SPR). Through a small telescope this region appears as a greyish region in the far south; however, even here there is activity, and larger telescopes reveal the existence of further faint belts and zones. In July 2009, an asteroid or comet impacted with Jupiter. The SPR bore a dark scar from the impact which was visible in large amateur telescopes.

OBSERVING JUPITER

As seen from the surface of the Earth, the path the Sun, Moon and planets make as they move across the skies is called the *ecliptic*. This line can be drawn on star charts, and it passes through the 12 constellations of the Zodiac (Taurus, Aries, Gemini and so on) and one further constellation which is not included in the Zodiac – Ophiuchus the Serpent Bearer. As Jupiter moves around the Sun in about 12 years, it spends about a year in each constellation. For northern hemisphere observers, this is wonderful when the planet is in the winter skies and reaches a great altitude, but it is not so good when the planet lies in the summer skies, where it is low down and disturbed by poor seeing (see Chapter 5, p.89).

Jupiter comes to opposition once a year, and this is the best time to observe it. Jupiter's vast size and striking cloud features means that even a 3-inch telescope will show quite a bit of detail on a good night. A 4-inch telescope will show the colourful cloud bands, and larger telescopes give truly stunning views of the planet; the dark storms and spots have delicate pastel shades and their movement is quite detectable after a few minutes.

Filters can be helpful when looking at Jupiter. As explained earlier, an optical filter simply allows some wavelengths of light to pass, and blocks others. You will find that a light blue filter really helps to bring out the details within the belts of Jupiter, while a red filter will help enhance bluish coloured features like the festoons of the equatorial zone. A yellow filter is a good all-rounder for just increasing the contrast between belts and zones.

THE GALILEAN SATELLITES

The four large moons of Jupiter – Io, Europa, Ganymede and Callisto – can all be seen with a small telescope, and it is delightful to watch their eternal cosmic dance as they pass back and forth around the planet. Through a small telescope, the moons appear as four star-like points, but through a 6-inch telescope or larger the moons look like small discs. An interesting event to watch out for is a *transit*, when one or more of the moons passes in front of the planet. When this happens, the moons cast shadows onto Jupiter's cloud tops. The inner satellites, Io and Europa, are quite close to

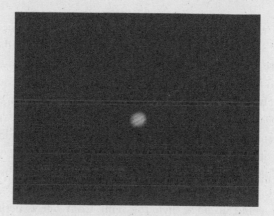

Jupiter and its four largest moons, imaged by Nick Damico.
From left to right they are Callisto, Europa, Io and Ganymede.

Jupiter and so their transits only last a few hours, but Ganymede and much rarer Callisto (as it is much further away from Jupiter) take longer to transit. Ganymede is the largest satellite of Jupiter (indeed it is larger than the planet Mercury) and it casts quite a large shadow on the planet. This can be seen with a 3-inch telescope.

Although there is now little scientific value in watching the Galileans, it is always nice to see them accompanying Jupiter in the sky, and remember the part they played in freeing us from Ptolemy's Earth-based cosmology.

THE SPEED OF LIGHT

The four Galilean moons may have helped prove Copernicus's point of view, but that was far from the end of their usefulness. In 1676, almost 70 years after their discovery, the Danish astronomer Ole Rømer used observations of Io to prove that light does not travel instantaneously from one place to another, but rather travels at a finite speed. For thousands of years, philosophers had debated the nature of light, with many assuming that it travels at infinite speed, taking no time at all to cross distances. This assumption was largely based on the fact that when you open your eyes, the stars appear immediately, so the light must therefore have traversed the immense distance instantly. Some stated that it could simply be very quick, with Galileo suggesting an experiment whereby two people standing miles apart tried to simultaneously open shutters on lamps. If light were to take, say, one second to travel the intervening distance then the light from one experimenter's lamp would take that time to reach the other's eyes, whereupon he would immediately open his lamp. The light from this second lamp would take a further second to travel back to the first experimenter, who would measure a two-second delay between the opening of his lamp and the apparent opening of the light from the second lamp. The experiment showed no time delay between these two events, and neither did a similar experiment involving timing the delay between a flash of lightning and the reflection from a distant hill or cloud. Of course, we know now that the reason these experiments failed to detect a delay is because light is so much faster than they could have imagined, covering a distance of one mile in just five millionths of a second – well beyond the ability of even the very best 17th-century physicists.

Realising that light might simply be too quick to detect, astronomers started looking for a similar experiment over larger distances. They tried looking at Earth's Moon, theorising that it should appear to enter an eclipse a short time after it actually enters the Earth's shadow, but also found no discernible difference – though since the interval is a matter of a couple of seconds this was also too short for them to measure. Ole Rømer's idea was based on a similar idea, but using eclipses of Jupiter's moons. In collaboration with French astronomers Jean Picard and Giovanni Cassini, Rømer observed eclipses of Io, the innermost of the Galilean moons. Io orbits Jupiter in a period of 42 and a half hours (just under two days), and regularly passes into the planet's shadow. Just like all other celestial bodies, the Galilean moons orbit with almost perfect regularity, always taking the same time to orbit Jupiter, and were seen as cosmic timekeepers by Rømer. But studying the combined observations, Rømer noticed something odd: the precise times at which Io entered or left Jupiter's shadow varied, appearing to occur either earlier or later than predicted by a matter of minutes – a difference easily measurable by astronomers of the time.

Rømer correctly postulated that the inconsistencies between the observations were due to the distance between the Earth and Jupiter varying as they moved around their orbits. He proposed that we can only observe these celestial events when the light has travelled from Io to Earth, and if light travels at a certain speed then as this distance varies over the course of a year, so the time it takes for us to observe the appearance or disappearance of Io can vary. The name for this effect was *mora lumina* (literally the 'delay of light') and was exactly what Galileo had predicted 30 years before, only with a much, much smaller effect.

Using his observations, Rømer correctly postulated that when Earth is closest to Jupiter we should see the eclipse happen slightly earlier than we would if it was further away. In September 1676, he predicted that the eclipse of Io on 9 November that year would appear to occur 10 minutes late, and this is exactly what he observed. Rømer calculated that light took 22 minutes to cross the Earth's orbit which, while a little slower than the true value we know today, provided the first ever measurement of this universal constant. Further work by such noted astronomers as Edmund Halley, Christiaan Huygens, John Flamsteed and Isaac Newton took into account the change in distance between the Earth and Jupiter over time. These calculations revised Rømer's value and led to the conclusion that light takes 17 minutes to travel from

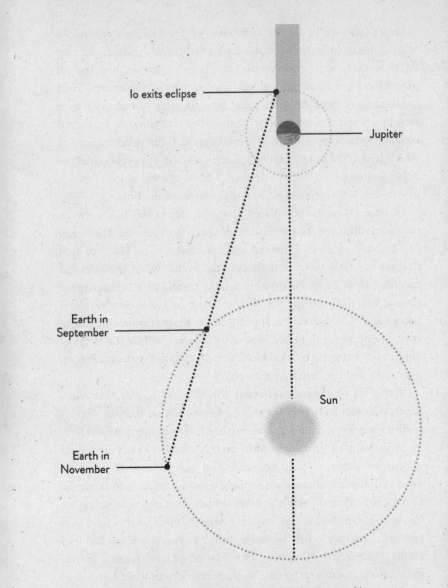

Io exits eclipse

Jupiter

Earth in September

Sun

Earth in November

Sketch of Rømer's observations of the eclipses of Io,
used to measure the speed of light in 1676.

the Earth's orbit, or eight and a half minutes to travel from the Sun to the Earth – not very far off the correct value of 8 minutes 20 seconds.

This value, based on Ole Rømer's measurements, would remain the best measurement of the speed of light until the 1720s, when James Bradley used the apparent positions of stars to calculate a similar value. Even then, all that was known was the time it took for light to cross the Earth's orbit, and the actual value of the speed was uncertain. Following observations of transits of Venus in the 1760s, which (as we have seen previously) allowed the scale of the Solar System to be measured, this speed was calculated for the first time. We now know that light travels at the staggering speed of 300,000 kilometres per second (186,000 miles per second), or 670 million miles per hour, and is the fastest speed that anything can reach.

CLOSE ENCOUNTERS

The exploration of the Jovian system began in the 1970s, initially with the Pioneer 10 spacecraft which flew past in 1973, followed a year later by Pioneer 11. These two probes were the first to cross the asteroid belt and gave us our first close-up views of Jupiter and (in Pioneer 11's case) Saturn. They were also the first probes to attempt a parts of a 'Grand Tour' of the outer Solar System, using the gravitational pull of the giant planets to slingshot them out of the Solar System so fast that they will never return. While the cameras on board the Pioneers were of limited quality, they led the way for further generations of missions to the outer Solar System. Just a few years later, in 1979, the encounters of the twin Voyager spacecraft with Jupiter truly revolutionised our view of this giant planet and its moons.

The Voyager 1 spacecraft passed within 300,000 km of Jupiter's cloud-tops, allowing astronomers to see features in the bands and zones that had never been seen before, some just 100 km in size or smaller. These included small white spots in the middle of the darker bands which, along with plumes of material on the boundaries between white zones and darker bands, seemed to rise above the darker material, spread out and finally fade as the material sank. By taking images over many hours, Voyager showed that the Great Red Spot is an enormous anticyclone (which just means that it spins anticlockwise), rotating roughly once every six days. Both this massive feature and the smaller white ovals are caused by the upwelling of material from deeper within the planet. The difference in colour between the Great Red Spot and the other features is still not fully understood – even to this day – but is thought to be due to the fact that the huge size of the storm allows it to pull up material from much greater depths.

In 1995, the Galileo spacecraft entered orbit around Jupiter after a six-year journey, providing us with the first in situ measurements of the Jovian system for a period of 8 years. At the very start of the mission, it released a probe which plunged into Jupiter's atmosphere and descended on parachutes to a depth of around 600 km below the cloud-tops before being crushed by the immense pressures. On the way down, it experienced wind speeds exceeding 600 km per hour (400 miles per hour) down to great depths, indicating that the source of energy driving the winds is not sunlight, as it is on the Earth, but Jupiter's own internal heat. The probe entered on the edge of a hotspot, which could explain why some of the measurements were different from what was expected. The probe detected the presence of ammonia clouds high in Jupiter's atmosphere but a distinct lack of water clouds, which had been expected based on measurements from Voyager.

The Galileo spacecraft continued to orbit Jupiter for eight years, by which point most of its instruments had failed due to the harsh radiation environment. To prevent an unintentional and uncontrolled collision with one of Jupiter's moons, the spacecraft was sent on a collision course with Jupiter and burned up in its atmosphere.

GALILEAN SURPRISES

It could be argued that the most fascinating discoveries from the Voyager and Galileo probes did not involve Jupiter itself, but its moons – particularly the four largest Galilean moons: Io, Europa, Ganymede and Callisto. The Voyager spacecraft showed that each of these bodies, little more than points of light as seen from Earth, is a unique world, hiding many secrets we are only now beginning to unravel. The innermost Galilean satellite, Io, has an incredibly colourful surface, pockmarked by craters. But a careful look at the Voyager 1 images showed that these were not impact craters, but volcanoes. Furthermore the Voyager images, which had a resolution as low as 1 km per pixel when the spacecraft was at its closest, showed that there were no impact craters detectable on the surface, meaning that it must be very young, possibly less than a million years old. This is not to say that Io formed recently, but that the volcanism is so widespread that it completely resurfaces the planet on a timescale of a million years or less. Both Voyager probes spotted the plumes of volcanic material spewing hundreds of kilometres into space, making Io the only object in the Solar System besides Earth to have active volcanoes on its surface. A comparison of the two missions, which flew past Jupiter four months apart, revealed that some of these volcanoes were still active four months after they were first seen, and over the course of the Galileo mission hundreds of active volcanoes were identified, making Io the most volcanically active body in the entire Solar System. Since Io is too small to retain any of its own internal heat, the cause of the volcanism must be an external one. As with all moons in the Solar System, Io is tidally locked to Jupiter, meaning that it constantly shows the same face. This would ordinarily mean that the tidal bulges are fixed in place, but in the case of Io the next two moons out, Europa and Ganymede, have a subtle effect. Europa orbits Jupiter at exactly half the rate of Io, and Ganymede at half the rate of Europa, and their resonance forces Io's orbit to be slightly elliptical, with the moon moving closer and further from Jupiter over the course of its journey round the giant planet. This motion causes the tidal forces, which thanks to Jupiter's immense mass are very strong, to vary and generate heat within the moon. This heat melts the material the rock is made of and causes the more volatile elements, primarily sulphur, to erupt onto the surface. It is this sulphur, and the many different compounds it makes, that give the surface its mottled appearance – not unlike a pizza.

This brings us on to the second of Jupiter's major moons, Europa. The same phenomenon that leads to volcanism on Io also keeps the interior of Europa slightly warm. The result is not as dramatic, but in some ways more intriguing, for reasons that should become apparent shortly. Voyager 1 did not make a close approach to Europa, but observations from Voyager 2 showed that its smooth, icy surface is criss-crossed by a network of dark lines. Europa's surface is fairly young, though nowhere near as young as Io's, implying that it has also been replaced over the course of the last hundred million years or so. The dark lines seen on the surface of Europa are cracks and fissures, which have formed when parts of the surface have melted and refrozen. In other regions, the surface has a much more complex pattern of lines overlaid by slabs of smoother ice, which has been dubbed 'chaotic terrain'. These cracks and fissures raised the possibility of there being a layer of liquid water beneath the ice, which fractures and allows the water to seep up from below. This theory was strongly supported

A volcanic eruption on Io, witnessed by the New Horizons probe.

by Galileo measurements of the way that Europa interacts with Jupiter's magnetic field, requiring the presence of a conductive, non-solid layer somewhere beneath the crust. This prospect of an ocean of water existing on one of the moons of Jupiter, albeit far beneath the surface, prompted great excitement. After all, on Earth the presence of liquid water invariably means life, and if the right conditions exist within Europa then perhaps life exists in the subsurface ocean. Given the low levels of energy present within Europa, any life there would probably be very primitive, possibly no more than microbes. If it were anything like life on Earth, such an ecosystem would require a source of sufficient chemicals, particularly carbon. These chemicals could be present if the ocean is in direct contact with the rocky interior of the moon, though this is less likely if there is another layer of ice separating the two. If there is life in this Europan ocean, however, getting to it will not be easy as the covering layer of ice is tens or hundreds of kilometres thick over most of the surface, although some regions show similarities with features on Earth where relatively thin ice lies over liquid water. There are regions that may well be thinner, and it may be possible one day to send a probe to melt through the ice and explore the ocean beneath, though such a mission is many decades in the future. But perhaps that isn't necessary to investigate the composition of Europa's ocean. More recent measurements from ground-based telescopes have shown that chemicals dissolved in the water, such as magnesium and sodium, seem to have been deposited on the surface, presumably from the sub-surface ocean. If water has leaked up from beneath and left these compounds behind as it evaporated then it may well have left other substances behind, and possibly even Europan microbes – if they exist.

The other two Galilean moons, Ganymede and Callisto, look more like what one might expect of a moon of Jupiter, being darker in colour and peppered with impact craters. These moons are larger than both Europa and Io, with Ganymede's diameter of over 5,000 km making it the largest moon in the Solar System – and even slightly larger than Mercury. The shape of the craters, appearing flatter than those seen in the inner Solar System, suggests that the outer layers of these moons are predominantly made of ice, though overall they contain roughly equal amounts of rock and ice. Callisto's more heavily cratered surface marks it out as the oldest, while Ganymede had some tectonic activity very early in its history which erased the earlier cratering. One of the most surprising discoveries of the Galileo mission was the detection of a magnetic field around Ganymede, generated within the satellite. The fact that there is a magnetic field implies that the moon has a relatively complex internal structure, with a molten iron core and possibly a

The Galileo spacecraft's view of Europa's scratched surface.

subsurface layer of ice or even water. This complex internal structure within Ganymede is not seen within Callisto, which appears to have only been partially differentiated before it cooled. The interaction of Callisto with Jupiter's magnetic field is not attributed to an iron core, but could possibly show that this moon also has a sub-surface layer of liquid water.

JOVIAN SMALL-FRY

Although the four Galilean satellites were the first ones discovered, they are far from alone in the Jovian system. By the mid 1970s, telescopes on Earth had already identified another ten moons, most much further out than the initial four, but one, Amalthea, was discovered to orbit closer to Jupiter than Io. In 1979, the two Voyager spacecraft discovered three more inner moons, orbiting closer than Io, bringing the total up to sixteen. In the meantime they'd even managed to lose one, with tiny Thermisto, originally spotted in 1975 and less than 10 km across, not being rediscovered until 2000. But there was nothing to rival the massive Galilean satellites, with the largest being just 200 km across – less than a tenth the diameter of Europa. While the remainder of the 20th century revealed no more, even with Galileo in orbit, by 2011 the total number of moons was 67. All of these new moons were discovered by ground-based telescopes and the majority by one team, led by Scott Sheppard of the Carnegie Institute for Science in Washington. While the eight innermost moons, including the four Galilean satellites, orbit within 2 million km of Jupiter, the others lie at distances between 7 and 30 million km. Most have very elliptical orbits which are tilted relative to Jupiter by between 10 and 60 degrees, and the outermost ones even orbit in the opposite direction. Combined with the fact that they are so tiny, almost all being less than 10 km across, it is likely that most of these are in fact captured asteroids. They are grouped together in several similar orbits, implying that some are even the result of a larger asteroid being broken up by tidal forces due to Jupiter's strong gravity.

Moons aren't all Jupiter has in orbit: as Voyager 1 observed some of Jupiter's inner satellites it detected the presence of a faint ring system. The rings are so faint and so close to the planet, lying within one Jupiter-diameter either side, that they are incredibly hard to observe from Earth, but were studied in detail by the Galileo spacecraft as it orbited Jupiter. They were also observed by Cassini in 2000 (marking the only time there have been more than two unrelated spacecraft simultaneously studying a planet other than Earth or Mars) on its way out to Saturn, and by New Horizons in 2007 as it swung past Jupiter en route to Pluto. The origin of the rings is the continuous bombardment of the inner moons by small meteorites, with the resulting debris spiralling in towards Jupiter.

Jupiter's immense gravitational pull turns it into a giant vacuum cleaner, sweeping up many small asteroids and comets when they stray too close. Some of these ended up as the outer moons of Jupiter, while others hit the inner moons and help replenish the ring system. But a few end up hitting the planet itself. The most famous of these was comet Shoemaker-Levy 9, which was only discovered in 1993 as a collection of 21 fragments, having been ripped apart during a very close encounter with Jupiter in 1992. The fragments ended up in a wide, looping orbit around Jupiter, and in July 1994 impacted with the planet over the course of six days.

The comet was discovered by Gene and Carolyn Shoemaker and David Levy, who were comet-hunting together using the 16-inch Schmidt-Cassegrain at Palomar, California. They discovered what looked like a 'squashed comet' on one of their photographic plates. Further analysis showed that the comet had passed by Jupiter which had broken it up into a series of fragments. Further calculations by the late Brian Marsden at the Central Bureau for Astronomical Telegrams showed when the comet, now dubbed 'Shoemaker-Levy 9', would crash into Jupiter. Over a period of a week in July 1994, astronomers watched for the first time as the fragments slammed into Jupiter, the largest fragment (fragment G) leaving a bruise on Jupiter the size of the Earth.

Although the impacts themselves occurred on the far side of the planet, and were not well seen from Earth, the Galileo spacecraft was just over 200 million km from Jupiter and in an ideal position to observe them. As the largest fragments of the comet hit the cloud-tops they released a huge amount of energy, creating fireballs several thousand kilometres high and reaching temperatures of thousands of degrees. What was unexpected, however, was the resulting cloud of dark material that hung high in Jupiter's atmosphere. As Jupiter rotated, the impact sites came into view from Earth, telescopes revealed massive black marks, some as large as the planet Earth, formed essentially of debris from the cometary material. As well as demonstrating the effect of a comet impact with a gas giant planet, the Shoemaker-Levy 9 impacts punched a hole in the upper layers of Jupiter's clouds and let astronomers peer beneath for a brief time. This allowed them to probe the lower cloud layers and revealed that there was much less water than expected, something that was confirmed the following year by the Galileo atmospheric probe. More recently, observations by the Herschel Space Observatory showed that Shoemaker-Levy 9 may well have deposited a reasonable quantity of water in Jupiter's atmosphere, with the expectation being that over years and decades this water vapour will gradually sink to greater depths.

Impacts such as Shoemaker-Levy 9 are rare, but it was far from unique. There is evidence on the moons Ganymede and Callisto that similar collections of fragments have struck them, in the form of straight lines of craters. More recently, with the advent of amateur astrophotography, additional impacts have been detected but only after they've occurred. On 19 July 2009, amateur astronomer Anthony Wesley discovered a dark impact scar in the clouds near Jupiter's south pole. This dark marking was very prominent and clearly visible with an 8-inch telescope. Smaller than the Shoemaker-Levy 9 impact, an analysis of the dark material revealed that it was probably an asteroid that had strayed to close. There have been more impacts seen since then, observed as small, brief flashes in Jupiter's atmosphere but not large enough to leave a scar. Since many astrophotographers take thousands of individual images per night, often throwing away all but the clearest pictures, there could be many other

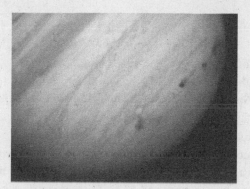

Black marks left in the clouds of Jupiter by fragments of comet Shoemaker-Levy 9.

impacts unseen in the huge archives of data. Jupiter's ability to sweep up comets and asteroids has been suggested as a factor in the development of life on Earth. By grabbing objects as they approach the inner Solar System, Jupiter could be protecting the inner planets from civilisation-destroying impacts. However, until we know more about the frequency of impact events on Jupiter it is hard to say whether this is true.

FUTURE MISSIONS TO JUPITER

The observations of the Jupiter system by visiting spacecraft have shown it to be a hugely interesting place. In 2011 NASA launched the Juno mission, which will reach Jupiter in 2016 and enter into a very close polar orbit around the planet. This will give us our first ever views of Jupiter's poles, and allow Juno to probe the interior of Jupiter, telling us finally whether it has a solid core deep beneath the clouds. It will also study the composition and structure of the cloud layers down to great depths, helping us understand how Jupiter formed and shedding light on the formation of the Solar System in general.

Although Juno is the only mission en route to Jupiter, the European Space Agency plan to launch JUICE (the JUpiter ICy moons Explorer) to study its moons in 2022. After flybys of Europa and Callisto, Juice will enter orbit around Ganymede, making it the first spacecraft to orbit a moon other than our own. The aim is to understand the surface and internal structure of this planet-sized moon, telling us how and where it generates its magnetic field and whether it has a sub-surface ocean. Although this mission will not arrive at Jupiter until the 2030s, the images and data it sends back should revolutionise our understanding of these icy moons. Its flybys of Europa will also measure the thickness of the ice for the first time, and select possible sites for a possible future mission to land on its surface.

10

Saturn

The planet Saturn is probably one of the most spectacular planets in the Solar System. A brief glimpse at it through a telescope reveals it to be a very different alien world to our own Earth. Saturn is a world of ice and gas; like Jupiter, it has an extensive atmosphere which gives way to hotter deeper layers of hydrogen and helium, forever hidden from our view. A look through Earthbound telescopes reveals only vague markings, but descend through the petrochemical smog which sits above the atmosphere, and we find winds and storms every bit as powerful and as terrifyingly large as those on Jupiter. Saturn is the only planet in the Solar System to have an enormous ring system visible to us here on Earth. Once thought to be solid, we now know it is composed of billions of fragments of rock and ice, all kept in place by a celestial ballet of moons and Saturn's powerful gravitational field.

Yet it is not only Saturn which has delivered some surprises. The satellites which the planet holds within its grasp have also proved to be quite exceptional. The idea that Saturn's moons would be largely dull lumps of ice has long been replaced by spectacular discoveries from the Cassini spacecraft of the icy geysers of Enceladus, and the thick atmosphere and rugged terrain of Titan.

Saturn continues to be of interest to amateur astronomers too. Its atmosphere produces occasional storms which are some of the most powerful the Solar System has ever seen. As we continue to explore the Saturnian system, we find treasures we never knew existed.

ANCIENT OBSERVATIONS

Saturn was the last of the five bright planets known to the ancients. Its great distance from the Earth means that Saturn appears to move slowly through the night skies. The Babylonians were some of the earliest astronomers to record the planet in their skies. A Babylonian observation from some time around 650 BC records an event whereby Saturn 'appeared to enter the Moon'. This event clearly describes an occultation of the planet by the Moon. Ptolemy in the ancient city of Alexandria examined Saturn at opposition and used his observation to help establish the details of the planet's orbit.

There are a number of mythologies featuring Saturn. To the Romans, Saturnus was the god of agriculture and plenty. The Romans even built a 'temple of Saturn' where the state treasury was housed. This magnificent temple stood on Capitoline Hill in the Roman forum, the ruins of which still exist today. Other cultures like the Chinese and Japanese associated Saturn as an *earth star*, and it was thus associated with one of the five elements from which, it was thought, everything was made.

We had to wait for the arrival of the telescopic era, however, before Saturn really started to reveal its surprises.

TELESCOPIC OBSERVATIONS

To see the rings of Saturn, you need a telescope of at least 2 inches. The early telescopic explorers of the Solar System were armed with instruments smaller than this, so one can imagine just how puzzling Saturn seemed; unresolved, the image in the eyepiece makes no sense at all. This was the situation which greeted the first telescopic explorer of Saturn, Galileo Galilei.

In July 1610, he turned his telescope of power ×32 towards Saturn. With the rings wide open (the angle the rings make to Earth change; see p.204), Galileo believed that Saturn was accompanied by two enormous satellites. Unlike the moons of Jupiter, however, these two enormous satellites did not appear to move and maintained both a constant size and position. The next time Galileo observed Saturn, in 1612, the rings would have been edgewise on to us, and Galileo observed that the two satellites had vanished and Saturn was alone. Galileo was to observe these mysterious attendants again as the rings opened out, and his drawings from the period show Saturn's rings in a number of guises. It seems that he never managed to work out what he was observing.

Galileo was not alone in his bewilderment. a drawing of Saturn by the French astronomer Pierre Gassendi also shows what appears to be a ring circling the planet. He too failed to reach any conclusions. This is not unreasonable – there was no reason to conjecture the existence of a magnificent structure like the rings of Saturn.

We had to wait for the arrival of the Dutch astronomer Christiaan Huygens to provide the right answer. Huygens was born at the Hague on 14 April 1629. His family was a wealthy one and this meant that Huygens was free to engage in intellectual pursuits. Indeed Huygens was something of a genius – not only did he invent the pendulum clock, he was also an excellent craftsman, grinding his own lenses and designing his own telescopes and eyepiece. His eyepiece design is now known as the Huygenian and is still popular today.

Some time during the spring of 1655, Huygens and his brother constructed a large aerial telescope some 12 feet in length. The telescope had a magnifying power of about ×50, and he began to study Saturn with the newly constructed telescope. One of the first things Huygens discovered was the existence of the largest moon of Saturn, Titan. On 25 March 1655, Huygens observed a small star to the west of Saturn. When he observed again, he discovered this small star had moved relative to Saturn. While the field stars behind Saturn came and went, this one single star followed Saturn in its path across the sky. The only explanation was that the faint star was a satellite.

Huygens stated himself that his discovery of Titan 'opened the way' to explaining Saturn's rings. He came to the conclusion that the object he and others had viewed was a thin flat ring which circled the planet without touching it. He correctly attributed its changing form to Saturn having an axial tilt. This in turn causes us here on Earth to view the rings from different angles – sometimes they are fully open; at other times they are edgewise on.

In 1675, the Italian astronomer Giovanni Cassini noticed that the rings were not uniform. He observed what appeared to be a thin dark line splitting Saturn's rings into two. This appeared in both telescopes he was using at the time – two long aerial telescopes 35 feet and 20 feet long. This dark line is called the Cassini Division and is a genuine gap between the rings, although spacecraft have since shown that the gap is far from empty, containing rocks and ice not visible to ground-based telescopes. Cassini went on to discover four satellites of Saturn: Iapetus, Rhea, Tethys and Dione. He also discovered that Iapetus is much brighter when on the western side of Saturn, and correctly attributed this to one hemisphere of the moon being darker than the other.

From 1774 to 1808, Frederick William Herschel made a number of investigations of the Saturnian system; indeed, he seems to have been truly captivated by the planet. In 1789, he discovered the two satellites Enceladus and Mimas using his 20-foot telescope.

Herschel was the first to note that Saturn is rather oblate – more so than Jupiter, in fact. He had also observed a number of faint belts on the planet, features which he took to be atmospheric in nature. In June 1780, he noticed that there was a rather dark spot within one of the belts. By following the spot, he estimated that Saturn rotated very quickly.

As telescopes improved, so more new discoveries trickled in. In 1850, the C-Ring was discovered by the American astronomers William and George Bond (William was George's father). William Lassell named the C-Ring the 'Crepe-Ring' since it appears to be slightly translucent. Where it passes in front of Saturn, it is possible to still see Saturn's globe shining through it. This was one of the pieces of evidence that showed that the rings are not solid objects, but composed of uncountable numbers of tiny icy particles following similar orbits.

The features of Saturn's atmosphere are rather more muted than those of Jupiter. Long-lived prominent storms like Jupiter's Great Red Spot are also absent from Saturn's atmosphere. As a result, finding an estimate for Saturn's rotational period proved to be difficult. Spots had been seen by both Herschel and Johann Hieronymus Schröter, but they seemed vague and elusive. On 8 December 1876, however, Asaph Hall was observing Saturn at the US Naval Observatory, which housed an excellent 26-inch refractor. Hall turned the great refractor towards Saturn and was rather surprised to see a bright white spot in Saturn's equatorial zone. The spot turned out to be rather long-lived, and this gave Hall – along with a number of American amateur astronomers – time to record how long it took the spot to traverse the disc. Hall then calculated from the observations an average rotational period for Saturn of 10 hours 14 minutes 23.8 seconds, a value fairly close to the modern-day value.

In 1903, the Spanish astronomer Josep Comas Solá observed that Titan was darker towards the limb (i.e. the edges of the disc) than in the centre. This is a phenomenon known as limb darkening,

Saturn's changing aspect over seasons
shown between 1996 (bottom) and 2000 (top).

and the effect is caused by an atmosphere. In 1944, Gerard Kuiper examined the spectrum of Titan and discovered it had an atmosphere consisting of methane. This was eventually confirmed by Voyager 1.

Today, Saturn is still very much of interest to both amateur and professional astronomers. For amateur astronomers, the main focus has been to monitor the planet for storms and outbreaks which normally occur every 30 years and seem to be seasonal (see p.211). The telescopes used by amateurs today are rather powerful, and when combined with photography, high-resolution images can be taken. This is invaluable since it seems the atmosphere of Saturn produces small short-lived storms which are frequently captured by imagers. As with Jupiter, the restless atmosphere of Saturn will keep amateurs and professionals busy for a long time to come.

THE CHANGING ASPECT OF SATURN

Saturn has an axial tilt of some 26.4 degrees. This means that, like the Earth, the planet experiences the seasons of winter, spring, summer and autumn. It also means that that the view we get of Saturn changes as Saturn moves around the Sun. The images opposite show Saturn's orbit around the Sun. When the rings are edgewise on, the Earth is in the same plane as the rings and so, in an Earthbound telescope, they appear as little more than a thin slither of light. As Saturn moves around in its orbit, so the southern hemisphere starts to tilt towards us, giving us excellent views of the southern hemisphere. The rings then start to close and we get a second edgewise presentation, after which the northern hemisphere starts to tilt towards us. The rings then close once more and we return to an edgewise presentation. The angle the rings make to us never exceeds 26.7 degrees in either direction, so the northern hemisphere never tilts more than 26.7 degrees towards us, and similarly the southern hemisphere never tilts more than 26.7 degrees to us.

The rings of Saturn are very bright, so the planet appears more brightly in the sky when they are fully open. When the rings are edgewise on, the planet appears fainter. The changing seasons do seem to have an effect on Saturn, and might well be responsible for occasional large storms which appear in the planet's atmosphere (see p.211).

THE FEATURES OF SATURN

A view of Saturn through a telescope shows that the planet has broadly the same sorts of features as Jupiter, although its belts and zones are not as well defined. This is not because Saturn's

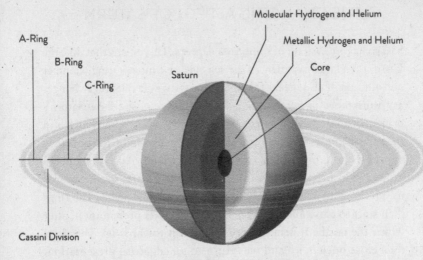

The features of Saturn and its main rings.

atmosphere is calmer than Jupiter's (indeed, Saturn's winds are much stronger), but rather is due to a form of petrochemical smog which hangs above the atmosphere in a most unhelpful way. As a result, only really prominent features can be seen.

The features which can be seen on Saturn depend largely on the planet's tilt. If the northern hemisphere is favoured, then the rings block our view of the southern hemisphere. Similarly, if the southern hemisphere is tilted towards us, the rings block the northern hemisphere. It is only when Saturn is edgewise (or nearly edgewise) to us that both the north and south hemispheres can be seen together.

In general, a 4-inch telescope will reveal the North and South Equatorial Belts when they are prominent. Larger telescopes will reveal the dusky polar regions, and the north and south temperate belts. Larger telescopes may well reveal the assistance of further belts and zones, but seeing conditions have to be good for this. Larger telescopes also bring out the pastel shades which are present at times. The North Tropical Zone and North Temperate Zones sometimes take on a slight greenish hue, and it is not uncommon for a bluish tint to be present in the north polar region.

Another phenomenon within reach of small telescopes is the wonderful jet-black shadow the ring system casts onto Saturn's cloud-tops. In a normal mirror-inverting telescope, where we have north at the bottom, after conjunction when Saturn appears in the morning sky, the shadow of the rings is below the ring system. As the apparition progresses, the shadow gets smaller and smaller until, at opposition, it all but disappears. After opposition, it reappears above the ring plane. The shadow of the rings on the globe provides a good test for seeing; if they are not jet black, the seeing conditions are rather poor.

For the purposes of amateur astronomy, we can split the rings into three. The outer ring is the A-Ring, which is separated from the B-Ring by the Cassini Division. After the B-Ring comes the innermost ring, the C-Ring.

When the rings are wide open and the seeing conditions are good, the A-Ring can be seen reasonably well through a 3- to 4-inch telescope. Larger telescopes show it to be darker than the B-Ring, and often a greyish colour. The B-Ring is the brightest of the three and normally white in colour. The C-Ring is rather dark, and greyish in colour. The best way to find the C-Ring is to look at the place where the rings cross in front of the planet. The C-Ring looks like a dark belt just below the B-Ring if the northern hemisphere is tilted towards us, or just above the B-Ring when the southern hemisphere is pointing towards us.

When the rings are edgewise on to us, because the rings are so thin, they seem to disappear in small telescopes, and Saturn then looks like a smaller version of Jupiter. Large telescopes may still show the rings, which look like a thin slither of light either side of the disc.

The globe of Saturn also casts a shadow on the rings. Just after conjunction (in a mirror-inverting telescope) the shadow of the globe on the rings appears on the proceeding side of the rings, just to the left of Saturn's globe. It then gets smaller until opposition, where it is absent, and then reappears on the right-hand side and increases once more until Saturn reaches conjunction.

OBSERVING SATURN

Saturn is a relatively easy planet to observe. It is fairly bright, and so it can be easily identified in the night sky. Even a small telescope will reveal a scene of staggering beauty – the small, pastel-coloured disc of Saturn surrounded by the magnificent, sparkling ring system, and a string of faithful satellite attendants shining faintly in the inky blankness of space.

A small telescope (3–5 inches, say) will reveal not only the rings and satellites, but also the darker equatorial belts which are located just above and below Saturn's equator. When the rings are wide open, the Cassini Division can be made out. Larger telescopes reveal more detail: a darker A-Ring, various details in the B-Ring and the translucent C-Ring. More belts and zones are also revealed, along with faint storms and irregularities in the belts.

Although a first look at Saturn discloses the obvious features, it actually takes a little time to see the more subtle cloud belts and ring structures, so don't be too disheartened if your first views don't reveal much in the way of fine details. Filters can help increase the contrast – a red filter will help bring out the bluish features, like the north polar cap. A blue filter will help to enhance the various belts on the disc. A yellow filter is a good all-rounder for helping to increase the contrast of features both on the planet and the rings.

Saturn's largest satellite is Titan, and this can be seen with binoculars. A 4-inch telescope should also show Tethys, Dione and Rhea. Larger telescopes will reveal Mimas and Enceladus, while very large apertures will also show Hyperion. The satellite Iapetus is an interesting one to follow: its two-toned surface means it is much brighter when on the western side of Saturn. If you have a small telescope, you will find that as Iapetus moves from the western side to the eastern side it becomes fainter and fainter until it can no longer be seen. When on the eastern side it has a magnitude of around +11 and requires a 6-inch telescope to see it. Once it moves back to the western side, it brightens and becomes detectable in smaller telescopes again.

STORMS

Because Saturn's atmosphere is just as active as Jupiter's it is not uncommon for storms to appear. Normally, small short-lived storms are recorded in the belts or zones by amateurs during the course of an apparition. Truly large storms seem to occur at 30-year intervals and are thought to be seasonal. There was the white spot observed by Asaph Hall in 1876. Another one appeared in 1903 and was observed by the famous astronomer of the time, Edward Barnard. An outbreak occurred in 1933 and was discovered by the British film and stage comedian Will Hay. Hay was using his 6-inch refractor at Norbury, when he came across the storm in very poor seeing conditions. Since then storms have occurred in 1960, 1990 and 1994.

These Great White Spots, as they are called, normally take the form of vast giant white ovals in the Northern Equatorial Zone. Even judging by Saturn's size, these storms are big and can extend in length to many thousands of kilometres. Initially, they are quite bright and can be seen in small-aperture telescopes. Eventually, the winds in the Saturnian atmosphere disrupt the oval and spread the material all the way around the equatorial zone causing it to brighten. Such storms can last for many months before all returns to normal once more.

In 2010, there was a rather unusual storm on Saturn. On 5 December 2010, the Cassini spacecraft detected an intense burst of lightning in the atmosphere of Saturn. Not long afterwards, amateur astronomers recorded a bright white spot in Saturn's North Tropical Zone. This white oval erupted into the most dynamic storm, perhaps the most dynamic storm ever seen on the planet. Over the course of the next few months, Saturn's strong winds spread the storm clouds eastwards until they ran all the way around the planet. At the time of writing, there are still relics of the havoc wreaked in the northern hemisphere. The energy from this storm heated the upper atmosphere in two locations, causing 'beacons' of infrared light which were observed by telescopes on Earth. Rather than fading from view, these two hotspots merged, creating a huge vortex larger than Jupiter's Great Red Spot, but much higher in the atmosphere than its Jovian equivalent. Saturn's vortex heated the atmosphere by 80 degrees relative to the surroundings, and gradually faded from view over a couple of years. The great storm of 2010 was entirely unpredicted, so it's worth keeping an eye on Saturn in case it throws up another unexpected storm.

THE RINGS OF SATURN

Of course, Saturn's most impressive features by far are its rings, with the main rings spanning almost 300,000 kilometres and stretching between 30,000 and 80,000 kilometres above Saturn's cloud tops. These structures have been known for hundreds of years, along with the Cassini Division between the A-Ring and the B-Ring. The entire system is in fact much larger than this, but these are the only rings that are directly observable from Earth. In the 20th century some fainter, narrower rings were found using the occultation method, which involves observing the brightness of a background star dip as the rings pass in front of it. Since the rings are not completely opaque, the star doesn't vanish completely, and the change in how much it dims allows the density of the rings to be measured. This led to the discovery in the 1960s and 1970s of the very faint D-Ring, the closest ring to Saturn, and the E-Ring, which at that time was the most distant. Such observations also proved that the rings are not solid bodies, but made of an uncountable number of tiny particles. The A-Ring and the B-Ring are predominantly composed of icy particles, typically ranging in size from a few millimetres to many metres across, which reflect much of the Sun's light back towards it, meaning that they appear light when seen in reflection from Earth. The darker regions, such as the C-Ring and the D-Ring, contain much finer material, which doesn't reflect light as easily and so makes them harder to see from here.

The full complexity of Saturn's rings wasn't appreciated until spacecraft flew past, first Pioneer 11 in 1979, followed by Voyagers 1 and 2 in 1980 and 1981. These first close-up observations confirmed the existence of the D-Ring and the E-Ring, and discovered the faint and narrow F-Ring and G-Ring. Part of the reason these rings were hard to discover from Earth is that they are made up of much smaller particles than the main rings. These tiny grains of dust scatter light in a forward direction, rather than back towards the Sun, and so are most easily seen from the far side of Saturn – something which is obviously only possible with spacecraft.

The Voyagers also showed that there is much more detailed structure in Saturn's rings than previously seen. For example the innermost D-Ring is actually three small 'ringlets', while the broader A, B and C-Rings contain a number of brighter ringlets and darker gaps. Even the dark Cassini Division is not completely empty, instead comprising bands of relatively dark material, similar in size to that which makes up the C-Ring, and separated by narrow gaps. The

Saturn and its rings seen backlit by the Sun by the Cassini spacecraft.

observations by the Voyager probes showed that most of these gaps and ringlets are due to the orbits of some of Saturn's moons, with the Cassini Division maintained by Mimas, the innermost major moon which orbits half as far out again. Other moons have a more direct impact, preventing the tiny particles that make them from straying too far from their current orbits and creating the finer structure seen by the visiting spacecraft. The Encke gap, for example, is kept clear by the tiny moon Pan, just 30 km across, while the narrow F-Ring is constrained by the slightly larger Prometheus and Pan, both around 100 km in size. Given their effect on the rings, these moons have become known as 'shepherd moons'.

This interplay was further explored by the Cassini probe, which arrived in orbit around Saturn in 2004 to begin its long-duration study of the ringed planet. Cassini uncovered truly intricate detail in the rings, giving close-up views of interactions between the rings and moons. The F-Ring was also shown to have a much more intricate structure, created by its shepherd moons Pandora and Prometheus and probably other much smaller bodies as well. Some of the intricate patterns in the main rings were also shown to be caused by tiny objects, some just hundreds of metres across. These larger objects may well have formed from the ring material itself. In 2009, Cassini observed the rings over the Saturnian equinox, when they were illuminated edge-on. It captured the shadows caused by vertical structures in the rings, demonstrating that in places the rings are just tens of metres thick – pretty thin given that they stretch for hundreds of thousands of kilometres. But there are larger vertical variations, such as wave-like motions seen on the edges of the Keeler gap, a 40-km-wide break near the outer edge of the A-Ring. These are caused by the gravitational pull of the tiny moon Daphnis, less than 10 km across, which orbits in a slightly tilted plane and so regularly moves above and below the rings.

Some of Saturn's moons have a much more intimate relationship with the rings, however. The Cassini images showed a few very fine rings associated with some of the newly discovered moons of Saturn, such as Anthe, Methone and Pallene, all of which are just kilometres in size. The origin of the material is thought to be impacts of micrometeorites with these moons, and the material is confined by the gravitational influence of the much larger Mimas which orbits just inside these three. The E-Ring is now known to be formed of particles spewing from the south polar region of Enceladus – of which (much) more later. And in 2009 an enormous ring was found by the Spitzer Space Telescope, more than twenty times further away from Saturn than the main rings and associated with the moon Phoebe. This ring is made of tiny particles which slowly spiral inwards towards Saturn and, like Phoebe itself, orbit in the opposite direction to the other large moons. This material is the favoured candidate for the origin of the two faces of Iapetus – again, more on that later.

There are some structures which are much harder to explain, though. As they passed by in the early 1980s, the Voyager spacecraft spotted bizarre-looking dark features that appearing to move around the B-Ring. These 'spokes' are thought to be caused by clouds of tiny particles rising above the ring plane, but their origin is much more mysterious. Since they seem to rotate around the planet at the same rate as its magnetic field, the favoured theory is that they are driven upwards by electrostatic forces, though they could also be caused by the impacts of micrometeorites with the rings, or the movement of larger objects within the rings themselves.

The biggest mystery of all about Saturn's rings, however, is how they got there in the first place, and even when they formed. They are predominantly made of ice, and all the giant planets have rings, but Saturn's are by far the most majestic. With the exception of the E-Ring and the 'Phoebe ring' they are close enough to Saturn that the planet's strong gravitation pull prevents them from forming a solid body, instead spreading out the material into a complete ring. This is why material thrown off some of the innermost moons forms rings, such as the F-Ring from Pandora and Prometheus, and the ring associated with Pallene. But the main A, B and C-Rings have no such obvious origin, particularly as they have a mass equivalent to a moon several hundred kilometres across. While any material this close to Saturn is prevented from forming a solid body, they are not close enough to be explained by a moon being ripped apart. If that much material had been so close to Saturn during its formation, it would all have been accreted into the planet itself, so it is generally assumed that the rings were formed after the family of moons. The gravitational pull of the moons could have prevented the ring material from falling inwards to Saturn, but it still has to be explained how so much ice and dust found itself in orbit around Saturn. One possible explanation is that the rings are the result of a massive impact, or impacts, which destroyed a moon at some point in Saturn's history, while another theory is that a massive moon, perhaps the size of Titan, was broken up by tidal forces and led to the formation of the rings and most of the large, inner satellites.

Since the rings are so bright, it is argued by some that they must be very young, as older material would have gradually picked up a fine coating of dust and become darker. Taking this line of thought further, it is argued that the darker material, such as that which makes up the C-Ring and the Cassini Division, may be older, having formed earlier than the brighter material in the A and B-Rings. The problem with this line of thought is that the rings are *really* bright, and if the brightness really is a measure of their youth, then they could be as young as a hundred million years. And if they only formed recently then they could disappear over a similar timescale, so we may simply be living at a special time. It seems unlikely, though not impossible, that massive collisions were taking place in the Solar System so recently, as the majority of such events took place during a period known as the 'Great Heavy Bombardment' around 4 billion years ago. If the rings are ancient, then there needs to be explanation for why they are quite so bright. Observations of the density of the ring particles made by the Cassini spacecraft showed that the rings are clumpier than previously thought, with the tiny particles sticking together, and this has been put forward as an explanation for their antiquity. In truth, there is a huge amount of debate about the origin of Saturn's rings, which is one of the many reasons why they really are one of the most wonderful features of the Solar System.

THE MOONS OF SATURN

If Saturn's rings are its greatest mystery, then its family of moons is arguably its second. Saturn has more than sixty moons and, like Jupiter's, the majority of them are tens of kilometres across or smaller, probably being captured asteroids or other minor planets. But while Jupiter's system is clearly dominated by the four massive Galilean satellites, Saturn's moons exhibit a much wider range of sizes and compositions. What they have in common is that they are largely made of ice, simply because they are far enough out in the Solar System that there was lots of hydrogen and oxygen around when they formed. The system is dominated by Titan, the second largest satellite in the Solar System at just over 5,000 km across, the only moon to have an appreciable atmosphere, and the only body in the Solar System besides Earth to have liquid on its surface. This all marks Titan out for special treatment, and we shall return to it in more detail later. There are six other major satellites, though all are considerably smaller than Titan.

Mimas, the innermost large satellite, is 400 km across and has the greatest effect on the rings and inner moons. As has been mentioned previously, some of the material in the Cassini Division, which separates the A and B-Rings, orbits at twice the rate of Mimas in what is known as a 2:1 orbital resonance. The inner edge of the B-Ring, which marks the boundary with the fainter C-Ring, is at the location of the 3:1 orbital resonance, meaning that particles there orbit Saturn three times for every once Mimas orbits. This means that Mimas's gravitational pull, while much weaker than Saturn's, repeatedly pulls on particles near these locations in the same direction every time, adjusting their orbits and causing the gaps to appear. A similar effect is thought to affect the orbits of other moons and rings, such as the G-Ring, and the tiny moon Pandora which 'shepherds' the outer edge of the F-Ring.

While all this resonating might make Mimas seem like a bully, it's certainly taken its fair share of pounding in the past. Its surface is heavily cratered all over, but dominated by the massive Herschel crater some 130 km across – around a third of the diameter of the moon itself. Had the object which created the Herschel crater been much larger, Mimas would probably have been broken up, depriving us of a Death Star replica in our very Solar System.

Saturn's moon Mimas, which is dominated
by the enormous Herschel crater.

Enceladus is a similar size to Mimas, but seems to have experienced a very different history. The images from Voyager 2 showed that its surface is not very heavily cratered, implying that it is geologically active and replacing its surface on a timescale of a hundred million years or so. This was supported by the presence of immense fissures over the surface, which could have been formed by tectonic activity, and the fact that Enceladus orbits near the densest part of the E-Ring, and so could be responsible for the origin of that material. When Cassini entered the Saturn system it identified that Enceladus is significantly denser than the other moons, implying that it may have a much larger rocky core. The radioactive decay of some elements in that core would produce internal heat and, combined with a tidal resonance with the larger moon Dione, could explain the origin of the geological activity. But the real surprise came when Cassini spotted material spewing from fissures near the south pole of this tiny moon. These 'fountains of Enceladus' are made of water vapour rich in dissolved minerals, indicating that they originate from a sub-surface ocean, which, given the high mineral content, must be a salty ocean. Amazingly, Cassini even flew through the plume and detected the presence of hydrocarbons which, when combined with the salty ocean and an internal heat source, makes Enceladus one of the most exciting places to explore in the search for life beyond Earth. But one big unknown is how long the fountains have been active: if they are a very recent occurrence then the formation of life seems less likely. The rate of flow of the water vapour is not constant, and has been observed to vary slightly over the course of several months, and so there is the possibility that the fountains stop and start. If the flow rate remains roughly constant, then estimates of the amount of material in the E-Ring imply that it cannot stay there for longer than around ten years, as the orbits are unstable for such small particles. Unfortunately, this means that there is no history of the Enceladus plumes recorded in the E-Ring, so to answer these questions we will need to study this tiny moon in much more detail.

The next three large moons are much bigger than Mimas and Enceladus, with Tethys and Dione measuring just over 1,000 km across while Rhea comes in at a diameter of 1,500 km – though this is still less than a third of the size of Titan. All three are heavily cratered but, as with most bodies of this size, show evidence of tectonic activity in their distant past. They all have fractures and chasms running around their surfaces, while Dione features ice cliffs hundreds of metres high, and a chasm stretches around three-quarters of the way around Tethys. Other than that, these seem to be fairly standard icy moons, made mostly of water ice mixed with small amounts of rock, but they have their subtle differences. While we can clearly see the overall shape of these moons and the structure and composition of their surfaces, showing them all to be covered in water ice, their internal structure can only be studied indirectly. One way of doing this is to measure their mass, based on how much of a gravitational tug they exert on a passing spacecraft, and so calculate their average density. Tethys, for example, has a

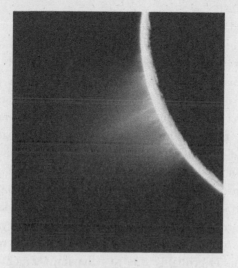

Plumes from the south pole of Enceladus.

density of 973 kg per cubic metre, which is only slightly higher than the density of water ice, which has a density of 930 kg/m³, and much, much lower than a density of 3,000 kg/m³ which is typical of rocky materials. The conclusion is that Tethys only contains about 6 per cent rock, significantly less than Dione and Rhea, icy moons which are made of between 25 and 50 per cent rock.

Further insight can be gathered by looking at the detailed shape, for no moon is a perfect sphere. Obviously there are craters and ridges, but ignoring those small-scale features the moons are in fact ellipsoidal, meaning that their diameter depends on the direction you measure it in – a bit like a rugby ball, or perhaps a walnut. Much of the asymmetry in their shape is due to the fact that all these satellites are tidally locked to Saturn, with one side always facing the planet, and Saturn's gravity stretches them slightly in its direction. The effect is small, typically between 3 and 20 km for these moons, so they are still pretty spherical, and in reality have nowhere near as extreme a shape as a rugby ball. The height of a moon's tidal bulges depends not only on the gravitational pull of Saturn, but also on how effectively the moon's own gravity pulls structures on its surface back down towards a perfectly spherical shape. This self-attraction depends not just on the total mass of the object, but also how uniform it is within. A completely homogeneous moon, with its material evenly distributed throughout, would have slightly weaker surface gravity, and therefore be able to support higher bulges on its surface than a moon which has an icy outer layer surrounding a much denser, rocky core. Detailed study of the shapes of these moons shows that they may have a different internal structure, and therefore possibly a different formation history. Dione, for example, is of a similar size to Tethys but has a much higher density. This is probably due to it having a much larger core, estimated to account for around half the total mass of the moon and to occupy around half its radius. The only other one of these large, icy moons to have such a high density, and therefore a larger rocky core, is Enceladus, which is slightly denser still than Dione. This may not be where the similarities end, and some of the latest results from Cassini indicate that there may be a very faint stream of particles flowing from Dione. This would be a much weaker equivalent of the activity on Enceladus, and is supported by the fact that the apparent flexing of the crust under some of the ancient tectonic ridges implies there may have once been a sub-surface ocean beneath the surface of Dione.

Both Dione and Tethys share their orbits around Saturn with two companions called 'Trojan moons', which follow the same orbit but a sixth of the way round, either ahead of or behind the main moon. Not discovered until the Voyager spacecraft flew past in the 1980s, these Trojan moons are much smaller than Dione and Tethys, with the largest being around 30 km across, and are likely have been 'trapped' in the Trojan points (see p.257) rather than formed there. As with all Saturnian moons they are named after characters in Greek mythology, going by the names of Telesto and Calypso (orbiting with Tethys) and Helene and Polydeuces (orbiting with Dione). Rhea may not have any orbital companions, but there were tentative signs in 2008 that it had a very thin ring system of its very own, based on an apparent lack of electrons in the vicinity. This would make it the first moon known to have rings, but sadly a thorough search for them in Cassini images over the following years showed nothing of the sort, and the initial detection may simply mean that the interaction of Rhea with Saturn's magnetic field is not understood. Although it probably doesn't have rings, Rhea does have the thinnest of atmospheres

The final large, icy moon of Saturn is Iapetus, with its distinctive two-tone colour scheme. Its discoverer, Giovanni Cassini, had noticed that its leading hemisphere, which we see when it is to the east of Saturn, appears to be only a tenth as bright as its trailing side. These suspicions were confirmed by the Voyager spacecraft, which also showed that the darker region is aligned almost perfectly with Iapetus's direction of travel. Although this could just be a coincidence, it does imply that whatever makes the surface dark is being picked up by Iapetus as it moves around its orbit, possibly from the more distant moon, Phoebe, which is not only dark but also orbits in the opposite direction to Iapetus. The discovery in 2009 of a ring of material orbiting along with Phoebe, and which gradually spirals inwards towards Saturn, added further weight to this theory. Unfortunately, Iapetus is so far away from Saturn, orbiting over three times further away than the other large moons, that the Cassini spacecraft has only had one close-up view of this enigmatic moon. This showed that the dark layer is a relatively thin covering, and that it must either have been ongoing for a long time or be a relatively recent event, since it covers the floors of all the craters. Apart from that insight, Cassini's observations only really deepened the mystery behind the source of Iapetus's dark side, showing that the material covering its leading face is of a different composition to the surface of Phoebe, being darker and redder in colour. The favoured theory at present is that material from Phoebe, rich in hydrocarbon molecules called 'tholins', does indeed impact Iapetus, but that something causes the composition of the material to change over time. These changes are likely to be driven either by the impact of micrometeorites, or the fact that Iapetus's long orbit means that over the course of its long day the weak sunlight causes the water ice to slowly migrate towards the poles, darkening the remaining material.

While we have better knowledge of what causes Iapetus's varying appearance, though by no means a complete understanding, the images from Cassini uncovered another big mystery: Iapetus is walnut-shaped. While most moons are fairly smooth ellipsoids, this one has a ridge running round its equator, which was certainly not expected of a moon 1,500 km in diameter. This ridge includes some of the highest peaks in the Solar System, rising over 20 km above the surface, and runs almost a third of the way around Iapetus – though oddly only through the dark, leading hemisphere of the moon. There is no firm explanation for how or why this ridge came to be, with possibilities such as the collapse of an ancient ring system or oddly concentrated tectonic activity all having difficulty explaining the existence of such a ridge. Since it is overlapped by craters, it must be an old structure, and also fairly solid. It may be related to the fact that Iapetus used to spin much faster, and have been caused by the moon getting tidally locked to Saturn, though no models can predict such a sharp feature. The fact that it only runs around the dark face of the moon, with the bright rear-ward facing side showing only a few isolated peaks, could even indicate that it is linked in some way to the two-tone colouration, though again it is far from clear how one process could manifest itself in two such different ways.

Hyperion, with its distinctive sponge-like appearance.

The remainder of Saturn's moons are typically much, much smaller, and many are likely to be asteroids that were captured into orbit around Saturn early in its history. Some, however, do deserve a brief mention. As has been mentioned previously, some of the moons are responsible for the structure in Saturn's rings. For example, Pan and Daphnis clear material from the Encke and Keeler gaps in the A-Ring, while Atlas patrols the edge of Saturn's main ring system. Slightly further out, Pandora and Prometheus keep the F-Ring in check, lying in very similar orbits and approaching within 1,500 km of each other on a regular basis. Slightly further out there are two small moons, Janus and Epithemeus, which practically share an orbit, and while they never get closer than around 10,000 km of each other they do switch positions every few years.

Hyperion has been mentioned briefly, and is the largest of the smaller moons at around 300 km across. Its odd shape, mostly due to a massive crater on one side, implies that it has been subjected to an enormous impact at some point in its history and may even be a piece of a larger object which was destroyed. The idea of a violent past is supported by its sponge-like appearance, with huge pits seen in its surface. Combined with its extremely low density, Hyperion may be little more than a pile of icy rubble stuck together by gravity, and coated by a dark material similar to that covering part of Iapetus.

The likely source of that dark material, Phoebe, is the most distant moon from Saturn with a diameter of more than a few tens of kilometres. Since it orbits in the opposite direction to the other large moons, and in a plane which is tilted by almost 30 degrees to Saturn's equator, it is likely to be a captured object, possibly originating beyond the orbit of Neptune in the Kuiper Belt.

TITAN

Saturn's largest moon merits a special mention for a number of reasons. Firstly, it is the second largest moon in the Solar System (after Jupiter's largest moon Ganymede), more than three times larger than Rhea and even outsizing Mercury. Secondly, it is the only moon known to have anything more than the thinnest atmosphere, and in Titan's case this is one and a half times as thick as the Earth's on its surface. The atmosphere is thick with nitrogen and small amounts of methane, making it opaque to visible light and impenetrable to the cameras on board the Voyager probes. The presence of methane, which is easily destroyed by sunlight, in the atmosphere has long been thought to suggest that there must be oceans on Titan's surface. But rather than being made of water these Titanian seas would be composed primarily of ethane and methane, simple 'hydrocarbon' molecules made of hydrogen and carbon. Not only is the surface temperature of -180 degrees Celsius about right for these to be in liquid form,

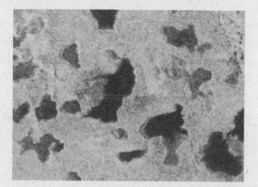

Titan's lakes, detected using radar measurements
from the Cassini spacecraft.

but without a supply of liquid methane on the surface the small
amounts in the atmosphere would be destroyed by sunlight.

In 2004, the Cassini spacecraft arrived in the Saturnian system
and released the Huygens lander, which descended through Titan's
thick atmosphere and made the first ever landing on a moon other
than ours. The images sent back as Huygens descended were
breathtaking, showing what looked to be a dried up lake-bed and
river channels, and even rocks exhibiting signs of erosion by either
wind or fluids.

Although Huygens showed only dried-up features, Cassini has
since made a number of flybys of Titan and confirmed the presence
of hydrocarbon lakes. For although the atmosphere is opaque to
visible light, infrared cameras can peer through the thick haze and
see the surface directly, showing that there is varied terrain over the
surface. While these images are not of high enough resolution to see
any lakes, radar measurements can show the texture of the surface in
much greater detail, demonstrating that some parts are so incredibly
smooth that they must be covered in a layer of liquid. These lakes
seem to be concentrated near the poles and can be rather large, with
many even given names. Ontario Lacus, for example, is around

200 km long and only slightly smaller than its terrestrial namesake, Lake Ontario on the US–Canadian border. It lies near Titan's south pole and over the course of several years was seen to drop in level, suggesting that it may be filled due to seasonal changes. Up at Titan's north pole, however, the lakes are much larger, with Kraken Mare stretching for over a thousand kilometres – almost as large as the Caspian Sea on Earth. While these lakes and seas contain liquid they are likely to be very different from those we are familiar with, with the main difference being that the liquid hydrocarbons are much more viscous than water, not unlike tar.

Titan's orbit is aligned with Saturn's equator, which is in turn tilted by about 26 degrees relative to its 30-year path around the Sun. Like almost all moons, Titan is tidally locked and keeps the same face pointed towards Saturn over the course of its 16-day orbit, but the axial tilt of the system means that it also experiences seasons. When Cassini arrived in 2004, it was late winter in the northern hemisphere and late summer in the south, and over the course of the mission we have been able to follow the change of seasons. Just as on Earth this has brought changes in the weather, with clouds on Titan seeming to form more readily in the winter. There has even been evidence of rain on Titan, with a storm near the moon's equator being followed by a brief darkening of the terrain beneath, suggesting that hydrocarbon rain dampened the surface for a short while. This rain would be made of orange droplets up to a centimetre in size, and possibly larger, falling as a slow drizzle due to Titan's weak gravity. Just like the water cycle on Earth, Titan seems to experience a hydrocarbon cycle, with methane and ethane in the lakes rising into the atmosphere, condensing as clouds, falling as rain and finally running over the surface into the lakes once again. As Titan shifts seasons into northern summer, Cassini will allow us to track the impact of the seasonal changes on its weather and climate.

Although Titan's surface is largely composed of frozen water ice, with a covering of hydrocarbons, its internal structure may be more complex than previously thought. Accurate measurements of the movement of Titan's surface imply that the icy crust is disconnected from the rocky interior of the moon, possibly separated by a sub-surface ocean. This ocean could lie as little as 100 km beneath the surface and be kept warm by the radioactive decay of elements in Titan's rocky core, combined with the presence of ammonia, which acts like antifreeze. The material from this slushy icy mantle might even make it onto the surface from time to time, with some evidence for 'cryovolcanism' – an analogue of volcanism on Earth but with icy materials rather than rocky ones. The direct evidence for cryovolcanoes is fairly weak, but such geological activity would also explain why there is quite so much methane in Titan's atmosphere, as even the seemingly large quantities in the lakes discovered by Cassini wouldn't be sufficient to maintain the observed atmospheric levels for more than around 10 million years.

The possibility of liquid water beneath the surface, combined with the presence of carbon-bearing molecules and a source of heat from the moon's interior makes Titan sound similar to moons we have met before, including Europa (orbiting Jupiter) and Enceladus (orbiting Saturn). Such conditions could possibly have led to the development of primitive life, though the very weak energy levels are far from favourable. It has also been suggested that life could survive in the lakes on Titan's surface, though with the lack of any oxygen it would have to use hydrogen in the atmosphere to generate energy. Again conditions are far from favourable, and the low levels of sunlight would make the pace of life very slow. Such theories of Titanian life forms are purely hypothetical at the moment, and not something that Cassini is likely to be able to detect. There are, however, plans for future missions, including a hot air balloon to explore the atmosphere and even a boat to sail on Titan's seas, though unfortunately none of these missions are funded at the moment.

CASSINI: LORD OF THE RINGS

We have discussed at length the discoveries that the Cassini mission has made about Saturn's rings and moons, but the spacecraft has also been studying the planet itself in great detail. It is true to say that we would not know anywhere near as much about Saturn if it were not for Cassini, which has shown us that this seemingly bland planet is much more complex than first meets the eye. The creamy hue that Saturn has as seen from Earth is due to high cloud layers, and when it first arrived in 2004 Cassini showed that the northern hemisphere actually had a blue tinge. Over the course of the following years the seasons changed on Saturn, moving from winter to spring in the southern hemisphere, and the creamy clouds returned. In the south, meanwhile, the blue hue started to appear with the onset of autumn.

Saturn also shows signs of more violent weather, though this is normally limited to relatively small white ovals, many of which have been monitored by astronomers on Earth. Being close-up, Cassini studied these storms in intricate detail, showing that Saturn experiences lightning a thousand times stronger than that experienced on Earth. It made detailed measurements of the 'superstorm' that erupted in late 2010, which was so very much larger than anything seen in recent times. Although this storm took astronomers, both amateur and professional, by surprise, historical observations indicate that large storms may well be seasonal effects. For some currently unknown reason they seem to have appeared more often in the northern hemisphere than in the south, so as Saturn heads towards summer in the north there's a chance that we may see similar events.

The wind speeds on Saturn are much faster than those on Jupiter, despite the fact that it receives less energy from the Sun, and have been observed in excess of 1,000 miles per hour. With no continents or land masses to stop them the winds spiral towards the poles, where

The hexagon around Saturn's north pole, with a
hurricane-like vortex in the centre.

massive vortices are seen to form as the cooler material drops down
into Saturn's lower atmosphere. At Saturn's south pole, an enormous
4,000-km diameter hurricane is seen, with the eye of the storm
allowing Cassini to peer to lower depths. Around this polar vortex, a
much more complex series of convection zones is seen. Such weather
patterns are observed in the atmospheres of other planets, such as
Venus, and even on Titan, but Saturn's is on a much larger scale.

If that was surprising, then there's something much more
astonishing going on around Saturn's north pole. The wind patterns
at a latitude of around 76 degrees cause a pattern of weather to
rotate around the pole in a hexagon shape. Such regular shapes
have never been observed in any other planetary weather system,
either on Saturn or elsewhere in the Solar System. Saturn's north
polar hexagon is thought to be the result of a coincidence between
the speed of the jet stream and the circumference of the planet at

that latitude. Laboratory experiments have shown that rotating weather systems can make a range of shapes if the wind speeds are just right, forming a regular pattern around the pole. The hexagon has even been imaged from Earth by amateur astronomers, which should allow its evolution to be tracked over the course of the summer, even long after Cassini's mission is over. Recently, as Saturn's north pole has emerged into sunlight for the first time in 15 years, Cassini has been able to study it in more detail and has even seen a massive hurricane right at its centre. This is a much more complex feature than at the south pole, being almost twice the size and with walls of cloud dozens of kilometres in height surrounding the eye of the storm.

Cassini has been in orbit around Saturn since 2004, and has seen the seasons on Saturn change over the course of a decade. It has made remarkable discoveries about the planet, its magnificent rings and its amazing family of moons. But no spacecraft can last for ever, and eventually Cassini will either fail or run out of fuel, risking an unintentional collision with Titan, Enceladus, or one of the many other moons. To prevent the possible contamination of these moons by bacteria from Earth, the mission is due to run until 2017, when Cassini will chart a course that will take it within the orbit of the rings before plummeting into the atmosphere of Saturn. The views as it does so should be stunning, but it will mark the end of what has been one of the most successful and revolutionary missions in the history of unmanned space exploration.

11

Uranus and Neptune

As we move further out into the Solar System, we come to yet another different type of world. Behind us, almost lost in the glare of the Sun, are the small rocky terrestrial planets. Beyond that, the Solar System is dominated by the two massive gas giants, Jupiter and Saturn. Ahead, in the gloomy, icy depths of the Solar System, lie the ice giants of Uranus and Neptune.

For most of its existence, mankind believed that there were only five planets in our Solar System. The telescope revealed not only the diverse nature of the planets in our Solar System, but also that it extended much further than had been realised. While the discovery of Uranus was a major achievement for astronomy, the discovery of Neptune was a major achievement for mathematics and the mathematical scientists who were able to predict both its existence and position simply by examining the orbit of Uranus.

The ice giants are a long way off, and to this day only one spacecraft – Voyager 2 – has ever visited them. Current research is carried out from ground-based telescopes and the Hubble Space Telescope. Amateur astronomers seem to have neglected these worlds once the initial excitement of their discovery passed. The ice giants may be remote, but they have their own important story to add to our understanding of the Solar System.

DISCOVERY: A NEW WORLD IN THE HEAVENS

Until March 1781, it was believed that the Solar System was complete. The five bright planets of antiquity were now known to orbit the Sun due to the recently discovered force of gravity. The new telescopes had revealed a great deal about our planetary neighbours; each of them had their own unique distinguishing features, and no doubt much more would be discovered about these worlds as telescopes improved in quality and size. The telescope was about to reveal much more than just new surface details on the known worlds. It would shortly reveal the existence of worlds and extend the boundaries of the Solar System far beyond imagination.

Friedrich Wilhelm Herschel was born in Hanover on 15 November 1738. His father was an oboist in the Hanover military band. At the time, Hanover was part of the United Kingdom and protected by George II. The war with the French eventually made its way to Hanover and, in the aftermath of the battle of Hastenback on 26 July 1757, the Hanoverian Guard sent by George II to protect Hanover was defeated. Herschel's father managed to get both Friedrich and his brother to the sanctuary of England.

Once in England, it seems that the young Herschel mastered the English language very quickly. He soon anglicised his name to Frederick William Herschel. He also seems to have inherited his father's musical talents: Herschel played the oboe along with the harpsichord and the violin. He worked as a musician at a number of places in England before finally settling in the town of Bath at 19 New King Street, becoming the organist at the Octagon Chapel.

Herschel had always had an interest in astronomy and, once at Bath, he began to take an interest in optics. He was an excellent craftsman; he made all of his own mirrors and lenses, patiently grinding the mirror into the correct shape. He seems to have been somewhat fanatical about mirror-grinding, and he would sometimes grind mirrors for up to 16 hours at a time, his sister Caroline feeding him and playing him music while he worked at his mirrors. It is not surprising that his optics and telescopes were much sought after by the most serious astronomers of Britain and Europe.

By May 1779, Herschel had made himself a fine 6-inch Newtonian reflector. (There is a replica of this at the Herschel museum in 19 New King Street, and *The Sky at Night* has filmed there on a number of occasions.) Herschel's initial interest seems to have been in double stars. These are stars which appear close together. They may be true binaries, trapped in each other's gravitational embrace, or they may be optical doubles and just happen to lie in the same part of the sky but have no physical connection. By October 1779, he was actively engaged in searching for new double stars, recording his findings in his log book.

On 13 March 1781, Herschel was busy examining the stars of Taurus when he happened upon an unusual object near the star ζ Tauri (zeta-Tauri). The nebulous object appeared to show a small disc – something no star should ever do. A subsequent search showed this unusual visitor to Taurus had moved position, indicating that it was within the confines of our own Solar System.

Herschel initially believed he had discovered a comet, and sent his results to the Royal Society. The Russian astronomer Anders Johan Lexell was the first to work out the orbit of the new path, and when he did this he discovered it was almost circular – a sure sign that the new object was a planet rather than a comet.

Other astronomers observed it and could find no hint of a coma or tail, and it seems by 1783 everyone was agreed that the object in question was a new planet, the seventh in our Solar System. King George III seems to have been particularly impressed with the discovery; he invited Herschel to become the King's Astronomer (this is different from the position of Astronomer Royal) with an attractive stipend of 200 pounds per year. The only condition was that Herschel had to move to Windsor.

Herschel originally wanted to name this new planet 'Georgium Sidus' (George's Star) after his royal patron. Unsurprisingly, this name was not popular outside England, and in the end the German astronomer Johann Elert Bode proposed the name Uranus, a Latin version of Ouranus, who was the father of Saturn in Greek mythology.

It became clear from observations of Uranus that there was something peculiar about it. On 11 January 1787, Herschel discovered two satellites, Titania and Oberon. By watching these moons circling Uranus, it became apparent that the axial tilt of the planet was quite pronounced – more than a right angle in fact.

Other astronomers observed the planet, but the small disc size made observations difficult. There was some disagreement over the nature of the surface features. William Lassell believed he had seen bright spots on the disc in 1892, but Lord Rosse was unable to see anything but a bland disc when he observed the planet with his enormous telescope in 1863.

DISCOVERY: ANOTHER NEW WORLD

Herschel was observing Uranus on 22 February 1789 when he noticed what he believed to be a ring circling the planet which was of a reddish colour. It is interesting to note that Herschel's drawing at the time does correctly match the size and orientation of Uranus's V-Ring. Observations from the Keck telescope have also shown that this ring is reddish in colour. This seems a genuine mystery since we know that the rings of Uranus are very dark, and subsequent observation by Herschel showed that the ring he believed he observed had vanished. No further sightings of a ring around Uranus were reported until the 1970s.

As astronomers got to grips with the new planet in their skies, so they began to observe and record its location. It wasn't long before a new problem arose. Lexell had used the recorded positions to establish Uranus's orbit and hence predict its location in the sky. However, Uranus seemed reluctant to obey Newton's laws of gravity – the planet seemed to deviate from where it should have been. Alexis Bouvard published a table of Uranus's movements in 1821, and suggested that the reason for the planet's unusual motion was the effect of some unseen body.

John Couch Adams was an undergraduate at Cambridge at the time, and he believed he could calculate the mass of this unseen planet using the existing orbital data. By September 1845, Adams had made some preliminary calculations and discussed the matter with then Astronomer Royal, George Airy. Although interested, it seems Airy believed a great deal of work would be needed in tracking down the new planet should it prove to exist. Meanwhile, the French mathematician Urbain Le Verrier had also noticed the discrepancies in the orbit of Uranus, and on 10 November 1845 he presented a paper to the Academy of Science in Paris about this very subject. Le Verrier was unaware of Adams's work and so his discovery was entirely independent.

When Airy heard of Le Verrier's paper, he quickly called a meeting of British astronomers, and it was suggested that James Challis, then director of the Cambridge Observatory, look for the new planet. Alas, a number of errors hampered their efforts: Adams revised his calculations but they contained errors, and Challis searched the wrong part of the sky. Worse, it seems that Challis recorded the planet twice; once on 8 August and again on 12 August, but he did not have the most recent star chart of the area, and so he did not realise he had recorded the new planet.

The search for the new planet was also hotting up in Europe. Le Verrier had been unable to convince French astronomers to look for his new planet but, undeterred, he wrote a letter to Johann Galle, director of the Berlin Observatory, asking him to search for the new planet near the position he had calculated. Although Adams had made a number of calculations to determine where the new planet was, there can be no doubt that Le Verrier had made his own calculations independently. His letter to Galle secured the discovery of the new world. On 23 September 1846, Galle and his assistant Heinrich d'Arrest began the search and, early on 24 September 1846, they found Neptune close to where Le Verrier had predicted. Galle and d'Arrest observed the planet for a few more nights and established its movements before announcing their discovery. The eighth member of the Solar System had been found.

The new planet had to be named, and it was Le Verrier who suggested the name Neptune, the god of the Sea. Friedrich Von Struve, another famous German astronomer was also in favour of the name Neptune, and so this became the name adopted.

Not long after the discovery of Neptune, William Lassell discovered the satellite Triton. Since then, observation of the planet has been somewhat scarce – this is not too surprising given the great distance of the planet. Even in the largest telescopes, the planet appears as a small featureless disc. Once more appearances would prove deceptive, as the Voyager 2 spacecraft revealed in the summer of 1989.

OBSERVING URANUS AND NEPTUNE

Both Uranus and Neptune can be seen in binoculars – the trick is knowing where to find them. Many astronomy magazines and books have charts showing their locations, and the determined observer will no doubt be able to track them down. Naturally, the best time to find these planets is when they are at opposition and so at their brightest.

Uranus has a distinctive greenish hue. A telescope of 3 inches or larger should show its small disc but that will be practically all. The surface features of Uranus are somewhat vague and elusive and nowhere near as well defined as the cloud markings of Jupiter and Saturn. Even large amateur telescopes only show slight hints of bands and zones. The satellites of Uranus will be invisible to small and moderate telescopes. Titania and Oberon may be glimpsed with an 8-inch telescope on a good night when the Moon is out of the way.

Neptune appears even smaller than Uranus, and has a strong blue colour. Even the largest amateur telescopes will reveal nothing of its surface. Its largest satellite, Triton, can be seen with an 8-inch telescope, though on excellent nights smaller telescopes may also show it.

Although the ice giants do not display any of the dynamic phenomena we see on the other planets, there is a joy in finding these worlds and looking at the outer bastions of our Solar System.

STRUCTURE AND ATMOSPHERE

Uranus and Neptune seem on the face of it to be very similar, with both being around four times the diameter of the Earth and around fifteen times the mass. They are both much bluer than the gas giants, with Uranus having a bright cyan colour while Neptune is more of a deep azure blue. The blueness is because of the absorption of red light by methane in their atmospheres, though the difference in colours implies a slightly different composition between the two. Most of the upper atmosphere of both planets is composed of hydrogen and helium, surrounding a mantle composed of water and ammonia. Although these types of compounds are often referred to as 'ices', the mantles of the ice giants are too hot and dense to be solid, and are likely to be dense, viscous fluids which smoothly change into gases. Between the mantle and the upper atmosphere are thin cloud layers of methane, ammonia and water ice crystals. This mantle contains the vast majority of the mass of the ice giant planets, while the very centre of them both is thought to be a dense, rocky core, probably with around the same mass as the Earth.

Neptune's axis is tilted by around 28 degrees to it orbit, meaning that it experiences seasons similar to those on the Earth – though with its 164-year orbit each one is much, much longer. Not only is Neptune incredibly distant and hard to observe but it was also only discovered just over one Neptunian year ago, so it will be a long while before we can try to understand its seasonal changes.

Uranus is very different. For some reason it is tipped over on its side by 97 degrees, so rather than moving round its orbit like a spinning top, Uranus rolls around. This makes identifying north and south a little difficult, but if you define the north pole as the one from above which the planet appears to rotate clockwise, then it is actually pointed just below the plane of

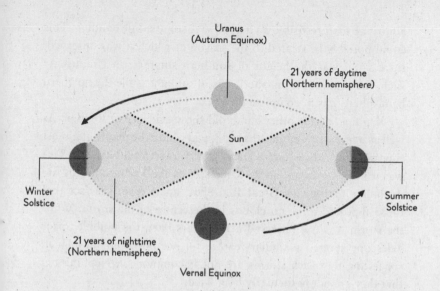

Uranus's seasons as it moves through its orbit.

its orbit. This has severe implications for its seasons, as in the northern summer its north pole is pointed almost directly towards the Sun, while its southern hemisphere is in perpetual darkness. 42 years later, the north pole is in darkness while the south pole is constantly illuminated. At the points in between, the planet's poles are pointing roughly along its direction of movement around its orbit, and the entire surface experiences a cycle of day and night much as the Earth does.

We have visited the two planets only once, with the Voyager 2 spacecraft passing Uranus in 1986 and Neptune three years later. Its sister craft, Voyager 1, had previously taken a course which gave it close-up views of Titan, Saturn's largest moon, but this directed it out of the plane of our Solar System and away from the remaining planets. The views Voyager 2 gave of Uranus and Neptune were strikingly different, which came as a surprise to astronomers. Uranus appeared as a smooth blue-green orb, with almost no texture visible on it at all save some subtle banding. This could possibly be related to the season at which it was observed, since 1986 was the height of southern summer on Uranus. As such, the southern hemisphere was in sunlight while the northern hemisphere was in complete darkness.

Neptune, on the other hand, did show evidence of weather. As well as a few wispy clouds that whipped round the planet at almost supersonic speeds, a 'Great Dark Spot' was seen on Neptune. This was a huge storm about the same size as the Earth, and not too dissimilar to Jupiter's Great Red Spot, apart from the fact that there appeared to be no clouds in the centre of the storm. While the storm was not observed a few years later, the Hubble Space Telescope spotted a similar feature further north on the planet, suggesting that such storms are not uncommon on Neptune but that they are not particularly long-lived.

There are also differences between the planets deep below the clouds, because Neptune emits much more heat from its interior than Uranus, despite the fact that it receives less than half as much solar energy. This internal heat dominates conditions on Neptune, driving the fastest recorded winds in the Solar System, with some seen to reach 2,400 km per hour. The winds on Uranus are much more sedate, though at up to 1,000 km per hour they are still extreme by Earth's standards. More recently, however, structures have been seen in Uranus's atmosphere by observing in near-infrared light, something that Voyager 2 was unable to do and which allows us to peer through the hazes to the action below. Observations from the Keck telescope, one of the largest on Earth, saw bright clouds and dark holes, possibly representing storms moving round the planet. Bright features near the north pole looked reminiscent of the complex convective zones found near Saturn's south pole, which could indicate that the weather systems are similar. At present the northern hemisphere of Uranus has just past the middle of

Neptune as seen by Voyager 2, showing the
Great Dark Spot and other storms.

spring, and as such the north pole is tilted only slightly towards the Sun. Observations over the last decade with ground-based telescopes have not seen any radical changes over the planet's equinox, as might be expected of a planet with such extreme seasons. Over the course of the next decade or so the north pole will come into better view, and we will get the first chance to look at what is going on in great detail. The decades following that will mark the coming of northern autumn and winter, and so by about 2070 we should have been able to observe a whole year on Uranus. It will be a long wait, but perhaps by then we'll understand how the orientation of this mysterious planet affects its climate and weather.

The Voyager 2 observations of the magnetic fields of Uranus and Neptune showed them to be very different from those of any of the other planets. While they are similar in general form, with both a north and south magnetic pole, they are tilted by over 45 degrees relative to the planet's axis of rotation, and centred at a location some way from the centre of the planet. This suggests that they are not generated in a liquid core, but possibly caused by the convection of a conductive layer within their mantle.

The lack of internal heat from Uranus is one of its biggest mysteries, and implies that at some point the interior of the planet lost its internal energy. This could be related to its extreme axial tilt, as theories have suggested that a collision with a more massive body could result in both odd properties. It is thought that both Uranus and Neptune have had a peculiar history, with some models of the formation of the Solar System indicating that Neptune formed closer to the Sun than Uranus. Early in their history the four giant planets would have been influenced by the gravitational pull of the others and by their interactions with the remains of the dusty disc of material from which they formed. Such interactions would cause them to move around in their orbits, with one possible result being that Uranus and Neptune ended up switching positions at some stage.

RINGS AND MOONS OF URANUS

William Herschel discovered the first two moons of Uranus in 1787, just a few years after he discovered the planet itself. It is a testimony to his abilities to both build and use telescopes that it was over 60 years before William Lassell discovered any more moons. These first four known moons were named in the 1850s by John Herschel, William's son, who, in a break with astronomical tradition, chose names from English literature. From William Shakespeare's *Midsummer Night's Dream* he picked Titania and Oberon, while from Alexander Pope he

Infrared images of Uranus from the Keck telescope.

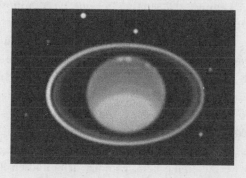

The rings and moons of Uranus.

picked Umbriel and Ariel. Since the fourth of these is also a character in Shakespeare's *Tempest*, when Gerard Kuiper discovered a fifth moon in the 1940s he chose the name Miranda from the same play. These five satellites are similar in many ways to Saturn's large icy satellites, with Titania and Oberon being a little over 1,500 km in diameter, Umbriel and Ariel being just over 1,000 km across, and Miranda being dwarfed somewhat at just under 500 km. It wasn't until Voyager 2 passed by in 1986 that any more were discovered, bringing the total number up to fifteen, and by 2003 this had risen to 27. These other satellites are all much smaller, ranging from 200 km in diameter down to less than 20 km. The Shakespearean naming scheme continued, using characters such as Desdemona from *Othello*, and Margaret from *Much Ado About Nothing*.

Far less is known about the Uranian moons than those of Saturn and Jupiter, largely because there has never been an orbiting satellite. The largest moons may well have liquid oceans beneath their icy crusts, as seems to be relatively common in large, icy moons. Miranda, meanwhile, experienced some severe resurfacing which created blocky structures on its surface. These large moons are thought to have formed from the same disc of material as Uranus; the smaller moons were almost certainly captured later on.

The rings of Uranus were discovered in 1977 by a number of groups of astronomers who were, at the time, intending to use the passage of Uranus in front of a background star to study its atmosphere. The discoverers saw five brief dips in the light from the background star, both before and after the passage of Uranus, and deduced that they must be due to a series of very narrow rings around the planet. They tentatively named them using the Greek alphabet, with the alpha, beta, gamma, delta and epsilon rings in order of distance from Uranus, and these names have stuck. This was the first time a ring system outside the Saturn system had been discovered, occurring a few years before Voyagers 1 and 2 would discover the rings of Jupiter. While they are far less spectacular than Saturn's magnificent ring system, it did show that such features are not unique in the Solar System.

With further analysis of the data from a number of teams, four additional rings were discovered, though the naming system became somewhat more confusing – they added an eta ring between the beta and gamma rings, and then three inner rings which were numbered 4, 5 and 6. A study of the locations of the rings compared to the orbits of Uranus's moons showed that, just as in the Saturn system,

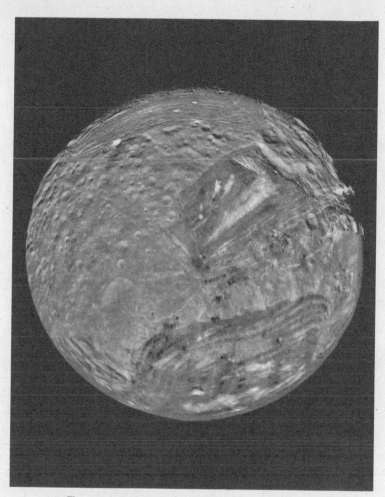

The patchwork surface of Uranus's moon Miranda.

the positions of the various rings are defined by gravitational interplay with the moons, in this case the inner ones. Of course, this method of studying rings only tells us anything about the tiny region which actually passes directly in front of the star, and it wasn't until Voyager 2 passed by that the first direct images of the rings came in. Along with the ten additional moons, these images also revealed the presence of two additional rings, later named the zeta and lambda rings.

In 2003, teams using the Hubble Space Telescope and the Keck telescope in Hawaii discovered a further two rings much further out than the existing rings and made of much fainter, finer material. One of these had a small moon embedded within it, Mab, which is probably the source of that ring material in a similar way to Saturn's E and G-Rings. As with Jupiter and Saturn's rings it is thought that interactions with the moons keep the rings in place, but with such small objects this cannot go on for very long. It is thought that either the rings of Uranus are relatively young, possibly of the order of hundreds of millions of years, or that they are constantly replenished by the collision and break-up of larger objects.

MOONS AND RINGS OF NEPTUNE

Being the furthest planet from the Sun, at a distance of more than 4 billion kilometres, there is little known about Neptune's rings and moons. Its largest moon, Triton, was discovered in 1846 by William Lassell, only a few weeks after the discovery of the planet itself. Triton is peculiar amongst large moons of the Solar System, as it orbits Neptune 'backwards' and at a tilt of more than 20 degrees to the planet's equator. This retrograde motion implies that Triton was not formed along with Neptune, as it would otherwise orbit in the same direction and be roughly aligned with the equator. Instead it is thought to have been captured at some point since Neptune's formation, probably having formed out in the Kuiper Belt. There have even been suggestions that it may originally have been one of a pair of objects which were separated from each other when they passed too close to Neptune.

Neptune's system of moons is somewhat different from those of the other giant planets, in that Triton is its only large satellite, with a diameter of 2,700 km. The other moons in the system are much smaller, and all discovered within the past century. It has been proposed that there may have been other large moons present when Neptune formed, but that they were destroyed or flung out of the system when Triton was captured. Gerard Kuiper discovered the second moon, called Nereid, in 1949, while the Voyager 2 spacecraft spotted six inner moons in regular orbits. Of these the largest is Proteus, at around 400 km in diameter. Compared with the Uranus system, the other inner moons are relatively large, with only one significantly below 100 km in size. In the early 21st century a further five outer moons were discovered, all in peculiar orbits. Some orbit in a retrograde motion, like Triton, but all of them orbit at angles which are tilted relative to Neptune's equator by up to 60 degrees. These orbits are all very eccentric, and make Neptune's system of moons the largest in the Solar System – in fact it's almost as large as Mercury's orbit around the Sun.

Since Triton is such a special case, Voyager 2 was sent on a trajectory that took it within about 40,000 kilometres of Triton. This allowed close-up views, but also meant that Neptune would be the last system that Voyager 2 could visit, as a journey past Triton sent it up out of the plane of the Solar System. As it passed by, however, measurements from the spacecraft confirmed its mass and size, showing that it is much denser than most of the other large moons in the outer Solar System. Its density it similar, in fact, to those of objects like Pluto and Charon, out in the Kuiper Belt, adding support to the theory that it was not originally formed with Neptune. Triton's surface, like its very tenuous atmosphere, is composed primarily of nitrogen and methane, with surface temperatures of around -230 Celsius. Its reddish colour implies that there are more complex organic chemicals on the surface. These chemicals are commonly called 'tholins', a term coined by Carl Sagan, and are thought to be produced as a result of sunlight breaking up the molecules of ammonia and methane, leading to the formation of larger molecules which settle on the surface. The size and shape of the cliffs and craters suggest, however, that the nitrogen, methane and tholins are a thin coating over a crust made primarily of water ice and ammonia.

The surface of Triton is relatively young, with very few craters, and shows signs of recent volcanic activity. The Voyager 2 images showed a bright south polar cap, covered in frozen nitrogen and methane, and a network of lines and fissures thought to be of tectonic origin. The south polar cap was also peppered with dark streaks of material, seemingly blown by the weak winds on Triton. It seemed very unlikely that such features could last for long on the surface, and so it was predicted that they are in fact very recent. A closer examination of the images revealed a few plumes of material rising around 8 km above Triton's surface before blowing downwind. These plumes are thought to contain compounds such as nitrogen gas, and are a type of cryovolcanism. The favoured theory is that sunlight heats up dark material encased in the transparent nitrogen ice on Triton's surface. The warmer particles of dark material eventually cause the nitrogen to vaporise, generating plumes which lift the darker material up into the atmosphere. This discovery of cryovolcanism on Triton leads to the possibility that other bodies out in the Kuiper Belt, which are thought to be of a similar composition, might show the same kinds of activity.

The rings of Neptune were the last system to be discovered in the outer planets, and were not discovered until the late

The faint rings of Neptune, seen against background stars.

1980s. They were eventually detected in a similar way to the rings of Uranus, by measuring the drop in brightness as they pass in front of a star. While there were dips which looked like rings in some observations, most showed no sign of them, and it seemed that if they were present then they were only partially complete. When Voyager 2 passed by Neptune in 1989 it confirmed that there were indeed complete rings, and detected five distinct rings in total. Some are broader and more diffuse than others but, as with the other giant planets, these rings seem to be associated with the innermost satellites, being shepherded by some of the small moons and probably created from material thrown off the moons by meteorite impacts. In contrast to the Solar System's other ring systems, however, Neptune's outer ring contains a sequence of dense arcs around part of its circumference. This clumpiness explains why only some ground-based observations detected the rings, as only the denser parts would have been observable by the experiments at the time. Such arcs would be expected to dissipate over the course of a few years, but a comparison with much more recent observations shows that they are largely unchanged. The exact reason for their confinement is unclear, but is most likely to be related to the 200-km moon Galatea, which orbits just inside.

While Neptune's moons were named after water gods from Greek and Roman mythology, the rings were named after astronomers who helped to discovered Neptune. These including the Galle ring, after Jonathan Galle, the first person to see Neptune, but also the Le Verrier and Adams rings, after Urbain Le Verrier and John Adams, who both independently predicted the position of Neptune.

* 12 *

Asteroids and Dwarf Planets

The major planets may be the dominant bodies in orbit around the Sun, but they only represent a small fraction of the total number of objects. They are joined by 'minor planets', which include the asteroids and dwarf planets. A small fraction, however, are found in orbits related to those of the major planets, particularly the gas giants, while others stray further towards the inner Solar System. Minor planets, as the name suggests, are much smaller than the major planets, and are the leftover pieces from the formation of the Solar System. The vast majority of them are very small, just tens of metres across or even smaller, but a few are hundreds of kilometres in size. Most of the minor planets found so far lie either in the asteroid belt, between the orbits of Mars and Jupiter, or in the Kuiper Belt, out beyond the orbit of Neptune. The largest minor planets are given a special status as dwarf planets. These include Pluto, which until 2006 was classed as a planet, and Ceres, the largest object in the asteroid belt.

These asteroids are of interest for a number of reasons. Firstly, some of them contain clues as to the very earliest stages of the Solar System's formation. Secondly, a few of them are on orbits which bring them very close to Earth, and our planet has been hit many times throughout its history. It is inevitable that this will happen at some point in the future, possibly with catastrophic consequences, and a better understanding of these objects is helping to develop plans for preventing such events.

The objects in the Kuiper Belt pose no significant danger to Earth, since they orbit at such a great distance, but their orbits can provide insight as to the history of our Solar System. Since these objects are very small and, in some cases, very distant, the majority are incredibly faint, and we still do not know where all of them are. The brightest are visible in binoculars, with Vesta just reaching naked-eye brightness from time to time. Their ongoing discovery is not just through the use of professional telescopes, but also through the efforts of amateur astronomers around the world. This makes minor planets one of the many fields of astronomy to which amateurs can contribute scientifically.

DISCOVERY OF THE ASTEROIDS

The location of the asteroid belt was predicted by astronomers in the 18th century, though perhaps for slightly dubious reasons. At that time the planets as far as Saturn had been discovered, and many astronomers were trying to find some sort of rhyme or reason behind their spacing. A number of astronomers, including Johann Titius from Germany, identified what seemed to be a logical sequence to their distances from the Sun. Since the scale of the Solar System had not yet been measured, the absolute distances of the planets were unknown, and as a result astronomers worked on the sizes of orbits relative to each other. This sequence worked fairly well as a numerical exercise and was subsequently publicised further by Johann Bode, and is now known as the Titius–Bode Law, or simply as Bode's Law. It gained further support several decades later, when William Herschel's discovery of Uranus seemed to continue the sequence beyond Saturn.

There was one glaring problem with this sequence of planets, in that for it to work there should be a planet between the orbits of Mars and Jupiter. To the astronomers at the time it seemed preposterous that a 'Lord architect' would have left that space empty. In the late 18th century, a group of astronomers from across Europe set about searching for this missing planet. In 1801, Italian astronomer Giuseppe Piazzi spotted an object which moved against the background stars from night to night. An analysis of the object's motion, combined with Kepler's laws of planetary motion, allowed him to calculate that it was on a nearly circular orbit which lies in the same plane as the other planets and puts it at just under three times the distance from the Sun as the Earth is. This new object, later called Ceres after the Roman goddess of agriculture, fitted almost exactly with the missing planet in the Titius–Bode Law, and as such was initially classified as the fifth planet from the Sun.

It should be emphasised at this point that Bode's Law is generally considered to be simply the result of coincidence combined with a natural human ability to see patterns – sometimes where none exists. The distances of the planets out to Uranus match the positions to within a few per cent, but Neptune broke the sequence when it was discovered in 1846. Besides, the general process of planet formation is fairly well understood, and there is no known reason for the planets to be arranged according to such a simple numerical rule. What is thought to be the case is that some of the planets and other objects are caught in resonances with each other, whereby there is a numerical relationship between their orbits, though this certainly doesn't apply in all cases.

The discovery of Ceres in 1801 was followed by the surprising discovery in 1802 of a second object at a similar distance from the

Sun, and over the next few years by a further two. All four were observed as being very small, with diameters less than the Moon's, although since they were so small the values calculated were rather inaccurate by today's standards. As a result, the objects became known as 'asteroids', meaning 'star-like'. By the 1860s there were dozens of asteroids, and while the first four – Ceres, Pallas, Juno and Vesta – retained special status, even they were eventually classed simply as asteroids. A naming convention for these first discoveries was eventually agreed upon and consists of a number followed by a name, with the number initially being assigned in order of discovery, and later by the order in which the orbit is accurately calculated.

The principle behind the discovery of asteroids is a relatively simple one, and hasn't really changed since Piazzi's discovery of Ceres in 1801. An observer images the same patch of sky at two different times, separated by minutes, hours or even days. Over such timescales, the stars stay absolutely stationary in the images, and therefore anything that moves between images is likely to be something in the Solar System. The key to defining an object's orbit is to track its location over many nights, and there are many astronomers, both amateur and professional, who dedicate much of their observing to this task. As well as individual observers, there are a number of networks of telescopes which observe large areas of the sky on a regular basis. The observations are collated by the Minor Planet Centre, which is responsible for monitoring the known asteroids on behalf of the International Astronomical Union.

Today there are more than 600,000 individual asteroids identified, around half of which have well-known orbits. The vast majority of these objects have been properly named, simply having a provisional designation – a code defined by the date of initial discovery, for example 1998 QE2. Of almost 18,000 named asteroids, many are named after mythical beasts and deities, though some are a little more whimsical. There is '9007 James Bond', for example, and '57424 Caelumnoctae' – Caelumnoctae is Latin for *Sky at Night*, which was first broadcast in 1957 on 24 April. There are even asteroids named after people, with examples including '4937 Lintott', '52665 Brianmay' and '2602 Moore'.

The vast majority of asteroids are found in the asteroid belt, between the orbits of Mars and Jupiter. This is not a completely uniform belt, with Jupiter's gravitational influence causing some regions to be almost devoid – much as some of Saturn's moons cause the gaps and divisions in its rings. But while most asteroids are found here, not all of them are. Some are found out between the orbits of Jupiter and Neptune, and are called Centaurs. Others share the orbit of Jupiter, sitting near points either 60 degrees ahead or behind the giant planet. These are the 'Trojan' asteroids and these locations are Jupiter's 'Trojan points', which are locations at which a smaller body can orbit safely. Similar objects have been found orbiting in similar points related to Uranus, Neptune and Mars, and NASA's WISE satellite recently discovered the first example of an Earth Trojan. While all these asteroids keep a safe difference, staying tens or hundreds of millions of kilometres from Earth, there are some which come significantly closer. These are the near-Earth asteroids, and we'll return to them later.

Despite the number of asteroids thought to be out there, they account for a very small fraction of the mass of the Solar System. The total mass of all the asteroids is thought to be about a twentieth of the mass of the Moon, and over a third of that is due to the few largest asteroids. It seems fairly certain that we have discovered all of the largest asteroids, with most of the unknown ones expected to be less than a kilometre in size.

There are very few things we can tell about asteroids from here on Earth, largely because they tend to be very small and (normally) very far away. As with the planets, they are normally seen by the sunlight they reflect. Because asteroids are generally very irregular shapes their brightness changes as they spin, and this variation can be used to monitor their rates of rotation. Most rotate in a matter of hours, though some are spinning as slowly as once every month or two. A few have even been observed to rotate in less than a minute. By studying the light reflected by asteroids in detail, such as through their spectrum, it is also possible to tell something about their composition. They fall into three main groups defined by whether they are predominantly made of rock, metal, or a mixture of the two. The rocky or stony asteroids are more correctly called carbonaceous since they are composed primarily of carbon. They are generally very dark, and account for around three-quarters of all known asteroids. The metal asteroids, generally referred to iron asteroids, and the mixed ones, called stony-irons, reflect much more light. As their names suggest, they are composed of a mixture of metals such as iron and nickel, with varying amounts of silicate materials. But why so many types of space rock? While all asteroids formed from roughly the same material the largest ones would have become differentiated, with the heaver elements, such as iron and nickel, sinking to their cores. It is thought that there were a lot of these large objects in the early Solar System, and that their break-up is responsible for the variety of asteroids we see today. Chunks from the core are composed largely of iron, while bits from the outside are much more rocky in composition.

METEORITES AND IMPACTS

Every now and again the Earth is hit by a small piece of space debris. The vast majority of the time these rocks are absolutely tiny and usually just burn up in the Earth's atmosphere. But, sometimes, they make it through the atmosphere and reach the

Meteor Crater in Arizona.

ground. If the object is more than a few tens of metres across, it is likely to cause a significant amount of local damage. For example, Meteor Crater in Arizona, which is just over a kilometre across, was the result of an impact around 50,000 years ago by an object which was probably only around 50 metres across. The reason for such a large crater is that the incoming object is travelling at tens of thousands of kilometres per hour, and so carries a huge amount of energy. The impact all but destroyed the original object, but the explosion as it hit the ground caused the vaporisation and ejection of the material from within the crater. Obviously, the larger the object the bigger the effect, and the impact of an object more than a kilometre or so wide would have global consequences, with the huge quantities of dust lifted into the atmosphere affecting the climate all over the planet. The impact which caused the extinction of the dinosaurs 65 million years ago was probably caused by an asteroid around 15 km across. One of the best pieces of evidence that the extinction was caused by an impact is the presence of a layer of rock laid down 65 million years ago which is incredibly rich in the element iridium. The Earth's crust does not naturally contain very much iridium, as most of it sank towards the core when the planet was molten, but it is much more common in asteroids.

While such impacts were much more common in the early Solar System, they are few and far between today. But much smaller objects, which do not leave a crater, do make it through the atmosphere and to the surface. These are called meteorites, and can range in size from grains of dust to massive boulders. As with asteroids, meteorites come in different types, with the three main ones being the stony meteorites, the iron meteorites, and finally the stony-irons. But unlike asteroids we can study meteorites in laboratories on Earth and so learn much more about them. The most interesting are the stony meteorites, also known as carbonaceous meteorites because they contain large amounts of carbon. There are two main sub-types of these which depend on their internal structure: the chondrites and achondrites. The chondrites are those that contain chondrules, which are round droplets of rocky material formed as the object cooled. These chondrules are easily destroyed if melted, meaning that these carbonaceous chondrite meteorites are lumps of rock that have never been particularly heated. As a result, they are the intact remains of the first stages of the Solar System, having been largely unaltered over the past 4.5 billion years.

Much of our understanding of asteroids comes from the study of meteorites, but not all of these rocks originate in the asteroid belt. Some come from much larger objects, blasted off the surface by a massive impact before travelling through space for millions or billions of years before landing here on Earth. The detailed composition of a few matches that of the moon rocks brought back by the Apollo astronauts, telling us that they came from the Moon. Similarly, others contain tiny pockets of gas, trapped in the rock during the impact which ejected them, and this gas matches the composition of the Martian atmosphere. Recently, analysis of one of these meteorites confirmed to a high significance that the minerals within it formed in the presence of water around 2 billion years ago.

ASTEROIDS UP CLOSE

Studying meteorites on Earth is incredibly valuable, but to learn more about asteroids themselves we have to visit them with spacecraft. This allows us to measure their masses and to study their surfaces in great detail. The first time this happened was in 1991, when the Galileo probe passed by asteroid 951 Gaspra on its way to Jupiter. It saw an irregularly shaped lump of rock about 20 kilometres long, covered in small craters and grooves. Judging by the apparent lack of large craters, it showed that Gaspra seems to be fairly young, probably a part of a larger asteroid that was broken up in a collision at some point in the last few hundred million years. In 1993, the Galileo spacecraft encountered its second asteroid, 243 Ida. This is a similar type to Gaspra but about twice the size, and was discovered to have its own moon. This moon, later dubbed Dactyl, is only 1.4 km across, and is thought to have been created in the same collision that created Ida itself.

We now know that about a sixth of asteroids are in binary pairings, sometimes composed of two objects of similar sizes, and sometimes with one object being much larger. We even know of some asteroids that have two moons. The types of collisions which create asteroids like Ida and Gaspra are thought to be relatively common, and it is predicted that there is a collision between moderate-sized asteroids about once per year. We've never see such a collision in action, but in 2010 the Hubble Space

The asteroid Ida and its tiny moon Dactyl,
discovered by the Galileo spacecraft.

Telescope took a picture of what is thought to be the result of two asteroids colliding about a year previously, creating a slowly expanding cloud of dust.

In 1996 NASA launched the NEAR-Shoemaker mission, which flew past 253 Mathilde in 1997. In contrast to Ida and Gaspra, Mathilde is a rocky asteroid, and given its apparently low density is probably made of a packed pile of rubble covered with a layer of dust. At around 60 km across, the rubble is held together by gravity, though large impact craters have shattered the surface. But Mathilde was only a waypoint along NEAR-Shoemaker's journey to the asteroid 433 Eros. The spacecraft passed Eros in 1998 before returning in 2000 and entering orbit, making it the first manmade object to orbit an asteroid. It went one better the following year, and after gently lowering itself to the surface made the first ever soft landing on an asteroid. Eros is a peanut-shaped asteroid, with a number of large craters. The largest of these appears to be responsible for the majority of the blocks of ejected material which litter the entire surface.

There have been a number of other missions to fly past asteroids, such as the Rosetta spacecraft which passed asteroids Steins and Lutetia. Its images of Lutetia showed that it was rather odd, having a slightly different appearance from most other asteroids. Comparisons with meteorites here on Earth showed that Lutetia doesn't belong in the asteroid belt at all, but probably formed much closer in to the Sun. The Chinese probe Chang'e 2 visited the near-Earth asteroid Toutatis as it made a close approach to Earth in 2012.

One of the most ambitious asteroid missions to date has been the Japanese Hayabusa mission. In 2005, this spacecraft travelled to the asteroid Itokawa – at just a few hundred metres in size, the smallest asteroid visited. Shaped like a potato, Itokawa shows very little sign of craters, instead appearing to be made of rubble covered in a thick layer of dust. Despite the tiny size of Itokawa, the Hayabusa spacecraft managed to control its position accurately enough to touch down on the surface. Following touchdown it managed to collect a few small samples of dust from the surface. Despite a number of problems throughout the mission, the Hayabusa team managed to bring the spacecraft back to rendezvous with Earth, and in 2010 the sample capsule plunged through the atmosphere and landed in Australia. The analysis of the particles, most of which were just a few thousandths of a millimetre in size, confirmed that the most common types of meteorites, a subset of the stony-iron meteorites known as ordinary chondrites, originate from stony-iron meteorites.

The latest mission to visit the asteroids is NASA's Dawn mission, which is visiting the two most massive asteroids, Ceres and Vesta. The aim of the Dawn mission is to help us understand the early Solar System by studying these two relics of its formation. It is thought that the inner planets formed from the collision of many smaller proto-planets, not unlike Vesta and Ceres. Such objects, ranging from a few hundred to a thousand kilometres across, would

Dawn's view of Vesta, the second largest object in the asteroid belt.

once have been much more plentiful, with these being two of just a few which are left over. Dawn arrived at Vesta in 2011, and spent just over a year orbiting the asteroid, mapping and analysing its surface. Although Vesta is largely a round shape around 500 km across, its south pole is dominated by a gigantic crater almost as wide as the asteroid itself. There is a family of smaller asteroids on a similar orbit to Vesta, which may well have been formed in the impact which created this crater, as well as the many other smaller ones which cover its surface. Some of those objects would have reached Earth, and it is thought that about one in twenty meteorites ever found came originally from Vesta. Being so large, the interior of Vesta is separated into a core, mantle and crust, in a similar way to the Earth and other inner planets. Studying the internal structure of Vesta can tell us what the objects that originally made up the Earth may well have been like, and provide clues about the very early Solar System.

Dawn will arrive at Ceres in early 2015, where it will become the first spacecraft to explore a dwarf planet in such detail. Ceres is around twice the size of Vesta, with a diameter of around 1,000 km. As a result, it is almost certainly differentiated, with a rocky core surrounded by a less massive crust. Observations from Earth have indicated that this crust may be partially composed of water ice, though we have to await Dawn's arrival to know whether that is the case or not.

COLLISION IMMINENT?

While asteroids are of interest for understanding the formation of the Solar System, particularly the inner planets, there is one group which attracts more attention than others. These are the near-Earth asteroids, which are on orbits which bring them close to Earth. The technical definition is those which come within about 200 million km of the Sun, which potentially brings them within about 50 million km of the Earth. A serious effort to search for these objects began in the mid 1990s, with projects such as Spacewatch and LINEAR. Lately, more advanced surveys such as the Catalina Sky Survey and Pan-STARRS have joined the hunt, and there are almost 10,000 near-Earth asteroids known, with that number increasing at a rate of almost a thousand per year. NASA's WISE telescope scanned the entire sky at infrared wavelengths, and made observations of hundreds of near-Earth asteroids. Its observations were compared with what is currently known about those asteroids, and allowed the team to predict how many asteroids of various sizes are out there. The conclusion was that there are around a thousand large near-Earth asteroids, with a diameter larger than 1 km, and that we have already identified more than 90 per cent of these. That's good news, although we have only found about 30 per cent of the mid-sized asteroids, with diameters above 100 metres across, of which there are expected to be about 20,000 in total.

Most of these asteroids spend most of their time outside the Earth's orbit, and so are not currently a risk, but chance encounters with planets or even other asteroids could change that by shifting their orbits. The most dangerous near-Earth asteroids, and therefore most carefully watched, are those which are known to come within about 7 million km of Earth's orbit and be estimated to have diameters of more than 150 metres. There are currently about 1,500 of these, and about a tenth of them are more than a kilometre across. These 'potentially hazardous asteroids' are carefully monitored, and their orbits predicted for up to a century in the future. The orbits are not known precisely, but there are none that have any real chance of hitting us at any point in the next hundred years or so. The closest approach of a potentially hazardous asteroid will be by Apophis, an object around a third of a kilometre across which will come within about 40,000 km in 2029. That encounter will adjust its orbit, and cause it to come close again in the future, but it is still not likely to hit us.

Of course while the objects we have discovered are not going to hit us, there is still a danger from those we don't know exist. One of the problems with finding near-Earth objects is that they spend a considerable portion of their time close to the Sun in the sky, and are therefore hard to observe. The tracking and monitoring of such asteroids does not only involve the large sky surveys, but also the dedication of a select group of amateur astronomers, who monitor the lists of new discoveries and help to pin down their exact orbits.

There is no denying that while an impact from an asteroid is unlikely to happen in any given year, it is inevitable that it will happen at some point in the future. That could be thousands of years in the future, or it could be next year. The point of monitoring the near-Earth asteroids is to give us enough warning when a large object is discovered to be on a collision course. The next step is to try to do something about it. While Hollywood might have us try to blow the object up, that is generally considered to be a bad idea. For a start, blowing up an object several kilometres across is very tricky, but even then we would be probably be left with many more smaller pieces impacting the Earth, doing almost as much damage as the large object.

Most of the work in collision-avoidance planning is looking at ways of adjusting the asteroid's orbit, and they sound like rather strange ideas. One solution might be to paint the asteroid white, so that it reflects more sunlight. The act of reflecting light causes a tiny push outwards from the Sun, and by adjusting the reflectivity the forces acting on the asteroid would be altered slightly. This would take many years to have an appreciable effect, but given enough time it could be sufficient.

Another way of adjusting an orbit is to give the asteroid a physical push. This would not be feasible using a chemical engine such as those which power rockets, but the ion drives that power spacecraft such as Dawn might be a solution. These use electric fields to accelerate tiny particles of ionised gas, normally xenon, and by sending them out of the back of the engine the spacecraft is pushed forwards a tiny bit. Each individual shove is very tiny, but these engines are incredibly efficient and a relatively small amount of fuel could power them for years. By attaching an ion drive to an asteroid it could be pushed into a different orbit, though there are likely to be complications caused by the asteroid's rotation.

The final method being considered is to 'park' a spacecraft close to the asteroid, and let its gravity pull the asteroid into a different orbit. The gravitational pull of a spacecraft is very small, but again, over many years the cumulative effect could be enough to adjust the asteroid's orbit enough to avoid a collision.

THE KUIPER BELT

The story of the discovery of the Kuiper Belt shares many similarities with the discovery of the asteroid belt. It began with a prediction, though rather than a numerical sequence this was based on the fact that Neptune didn't quite seem to follow the expected path around its orbit. In 1905, the astronomer Percival Lowell announced that Neptune was being affected by the pull of another planet. Neptune itself had been discovered in a similar way in 1846, based on oddities in the motion of Uranus (see Chapter 11). Percival Lowell searched in vain for a decade, but it was not until 1930 that the young astronomer Clyde Tombaugh spotted a faint dot moving slowly against the background stars. This object was called Pluto, a name chosen by an 11-year-old English schoolgirl called Venetia Burney, and was hailed as the ninth planet of the Solar System. It is a historical irony that Percival Lowell had actually captured Pluto in some of his plates but failed to notice it, possibly because it moved so very slowly against the background stars. It would also appear that Lowell's prediction that Pluto was causing Neptune to deviate from its orbit was incorrect, as it is far too small to have an appreciable effect, and the discovery is largely hailed as being down to good luck – combined of course with some very diligent observing on Tombaugh's part. Pluto is, however, affected by the presence of Neptune. The much higher mass of the ice giant means that for every three times it orbits the Sun, Pluto goes round twice – they are in a 3:2 resonance. This interaction also causes Pluto's orbit to be elongated, such that its distance from the Sun varies between 30 and 50 times the Earth–Sun distance, and also to be tilted at an angle of around 17 degrees to the orbits of the rest of the planets.

Over the next few decades, observations of Pluto showed that it had all the hallmarks of being covered in methane ice,

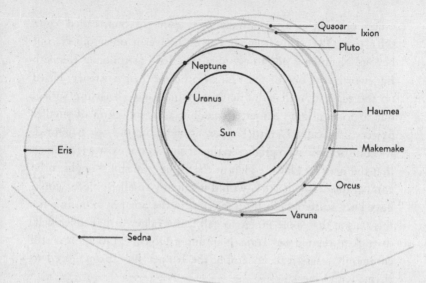

The orbits of some of the larger Kuiper Belt objects.

which is much more reflective than bare rock. This means that Pluto is much smaller and therefore less massive than the initial observations suggested. In 1978, the mass of Pluto was calculated much more precisely by the discovery of a moon, later named Charon. The observations showed that Charon is around half the size of Pluto, and therefore about one tenth of the mass, and that the two orbit each other in just over six days. When combined with their separation, this allows a measurement of the total mass of both objects, which was calculated to be about two thousandths of the mass of Earth. More recently, four much smaller satellites of Pluto have been discovered since 2005. Two are called Nix and Hydra; at the time of writing, the last two have yet to be named.

Pluto and its moons did not remain alone for long, and in 1992 astronomers David Jewitt and Jane Luu discovered another object orbiting at a similar distance from the Sun, called 1992 QB1 – or QB1 to its friends. The existence of other objects out beyond Neptune is not a new idea. The Irish astronomer Kenneth Edgeworth was one of the first to propose their existence, based on the suggestion that the material would have been too spread out to form a large planet, remaining as a collection of smaller objects. Ironically, Gerard Kuiper, after whom the belt is named, was one of those who predicted that it should not exist. He claimed that the mass of Pluto, which at that time was still thought to be comparable to the Earth's, would mean any smaller objects would have been scattered much further out in the Solar System. It is for this reason that some prefer to call it the Edgeworth–Kuiper Belt, or stick to the phrase 'Trans-Neptunian Objects', though it is still commonly referred to as simply the Kuiper Belt (pronounced to rhyme with Viper).

The discovery didn't end there, and objects started flooding in from teams around the world, with the number of Kuiper Belt objects now standing at over a thousand. Since they are so far away, the discovered objects are typically much larger than an average asteroid, with some being comparable to Pluto's 2,300-km diameter. Some of the more notable ones even have names. The rules of the International Astronomical Union state that Kuiper Belt objects have names relating to creation mythologies and, since most of the Greek and Roman names were taken, the Kuiper Belt objects have more international names. Haumea (How-may-ah) and Makemake (Mak-ee-mak-ee), for example, are named after deities from Hawai'i and Easter Island, and are about half the size of Pluto. Quaoar (Kwar-war), meanwhile, is named after a creator god of the indigenous Tongva people of south California. Like Pluto, these objects are all made primarily of ice, though they have varying amounts of methane and ammonia mixed in with it. Some of them are even known to have moons, such as Hi'iaka and Namaka orbiting Haumea, and Weywot orbiting Quaoar.

In 2003, a team led by Mike Brown found an object more distant than Pluto, orbiting between 40 and 100 times further from the Sun than the Earth. Now known as Eris, initial estimates made it out to be both slightly larger and slightly more massive than Pluto. With Pluto no longer being the largest in its class, this raised the question of whether Eris was the tenth planet in our Solar System, and also whether there were more out there. In 2006, the International Astronomical Union decided that Eris was not a planet and that neither was Pluto any more, giving each the status of 'dwarf planet'. This name applies to an object which orbits the Sun and is large enough to have become a spherical shape, but which is not massive enough to have cleared its orbit of other large objects. At present there are five dwarf planets in the Kuiper Belt: Pluto, Eris, Makemake and Haumea. These are joined by Ceres, the largest object in the asteroid belt, and currently the only dwarf planet in the inner Solar System. As telescopes get more advanced and surveys cover larger areas of the sky it is likely that we will discover more dwarf planets, and there are certainly dozens on the 'waiting list'. All that is required is confirmation of their size and mass, which would tell us whether they are round or not, though at such great distances such observations are very hard to make.

Although the objects discovered are a tiny fraction of the total number thought to be out there, it is already clear that they are arranged into groups. These groups are thought to be related to the formation of the Kuiper Belt and the influence of the giant planets, particularly Neptune. Theories of planetary formation suggest that Neptune formed closer in to the Sun than Uranus, but the gravitational pull of Jupiter caused it to move outwards in its orbit, ending up switching places with Uranus. The outer regions of the Solar System were full of small icy bodies, and as Neptune moved outward it would have had two main effects. Firstly, it would have scattered many of the objects much further out into the Solar System, forming a 'scattered disc' of objects way beyond Neptune. Many of these objects are in orbits which are highly elongated and often significantly tilted. Secondly, as it moved out it would have captured objects into resonant orbits, similar to the way in which groups of asteroids in the asteroid belt are affected by Jupiter. This includes a group of objects which are in the same type of orbit as Pluto, and are now classed as 'Plutinos', although there are a number of other resonant groups. Not all of the initial collection of Kuiper Belt objects would have been affected by Neptune, and the 'classical Kuiper Belt' is the group of objects in nearly circular orbits at a distance of 6 billion km or so from the Sun. One peculiar property of this classical belt is that it has a relatively sharp outer edge at a distance of about 7 billion km from the Sun. This cut-off of the disc could be related to the effect of Neptune as it moved out, or could just be a gap in the belt, with an outer Kuiper Belt existing beyond. Objects much more distant would be incredibly hard to detect with today's technology, and so we must wait and see whether the main Kuiper Belt continues.

We still know very little about these objects, but in a few years we should know more. The New Horizons probe is on its way to Pluto, and should get there in 2015. It will take pictures of Pluto, its large satellite Charon, and its other smaller moons, but is travelling too quickly to be able to enter orbit around any of them. After a nine-year journey, during which Pluto has been demoted from a planet to a dwarf planet, it will only have a few short hours to take all its measurements, before zooming onwards into the outer fringes of the Solar System. One of the questions it should be able to answer is what Pluto's thin atmosphere is like, and whether there is any sign of the cryovolcanism which has been seen on Neptune's moon Triton.

After visiting Pluto, it is hoped that New Horizons will be able to visit at least one more object while it is still able to communicate with Earth, and the team have been encouraging the search for objects along its flight path. Unfortunately, from our point of view on Earth, Pluto is in roughly the same direction as the centre of our Galaxy, and so the view is crammed full of stars. It must be hoped that by the time New Horizons leaves Pluto there is somewhere for it to go next. Even if nothing is found, it will be our first foray into the Kuiper Belt, and should provide invaluable information as to what the icy outer reaches of our Solar System are like.

BEYOND THE KUIPER BELT

There are large objects which stray much, much further out than those we have discussed. One such example is Sedna, an object around 1,500 km across which was discovered at a distance of 15 billion km from the Sun. Further observations showed it to be on an incredibly long orbit, taking over 11,000 years to complete a circuit and coming no closer than about 12 billion km. At its furthest, it is nearly a thousand times further from the Sun than the Earth. As a result, it is one of the coldest, most distant objects yet discovered, making its name all the more apt – Sedna is the Inuit goddess of the sea. Sedna's highly elliptical orbit implies that it has been perturbed at some point in its history, having started off somewhere near either its closest or furthest point from the Sun. One theory is that it started out much further out and was put on this orbit by the influence of a passing star. This might seem unlikely, but the Sun most likely formed in a group of stars, and billions of years ago they would all have been much closer together. It's even possible that Sedna started off in orbit around another star, and was captured by our Sun. That would make it the first interstellar visitor we've discovered.

Perhaps a more intriguing possibility is that Sedna was placed on its long, looping orbit by a planet orbiting dozens, or perhaps hundreds, of times further from the Sun than those we are familiar with. An object the size of Earth, or possibly even Jupiter, would be incredibly hard to detect out at those sorts of distances. It's even thought possible that the Sun has a faint companion of its own, a brown dwarf which is on the boundary between a massive star and a planet. Although these objects are incredibly hard to find, a careful search of the data from the WISE satellite, and from upcoming surveys by ground-based telescopes, may reveal its presence. One thing is for sure – the outer Solar System is far less boring than it seemed to be a few decades ago.

* 13 *

Comets

Every now and then, a ghostly visitor appears in the skies above. A bright comet in the night sky can be a glorious sight, a bright majestic tale sweeping back from the fuzzy nucleus. Once seen as omens of bad luck, we now know that comets are just another family of objects that exist within our Solar System. Though we no longer attach the same omens of doom and despair, it is not entirely true that comets pose us no risk. In 1994, astronomers watched as comet Shoemaker-Levy 9 smashed into the planet Jupiter, leaving bruises in the upper atmosphere the size of the Earth. There can be no doubt that comet impacts in the past have seriously damaged life on Earth.

Our understanding of comets has come a long way thanks to the observations of both professional and amateur astronomers and, of course, the spacecraft that have visited these icy travellers. Today, amateurs continue to examine the skies both for new comets, and to monitor known existing ones. Comets are erratic and unpredictable objects; they may brighten suddenly, and the brightness of new comets is notoriously difficult to predict.

In recent years, there have been a number of new discoveries, and it seems that comets may not only have been an essential part in delivering water to the Earth, they may also have been responsible for transporting here the basic ingredients of life itself.

EARLY HISTORICAL
ACCOUNTS AND MYTHOLOGY

It seems that comets have long been associated with bad omens and ill fortune. This has probably much to do with their sudden and frequent appearance in the skies above. Before the 'Celestial Revolution' of Brahe and Kepler, the external Universe was thought to be the realm of God; it was perfect and harmonious. Not surprisingly then, anything seen to disrupt that would portend bad luck for the hapless inhabitants of Earth, a sure sign that some god somewhere had been offended. Some went further, suggesting comets were attacks from the people of the heavens.

It is reasonable to suppose that prehistoric humans would have noticed the appearance of a bright comet in their unpolluted skies. As humans began to record their activities, we find descriptions of comets in the early Chinese oracle bones. Halley's comet is a periodic comet and returns to our skies every 75 years or so. It is fairly bright and can be easily seen with the naked eye. If we examine the historical records, we see this comet making an appearance far back in antiquity.

Perhaps one of the earliest recordings of the comet was in ancient Greece in 467 BC. In 87 BC, the Babylonians recorded the passage of Halley's comet in their stone tablets. The Chinese of the Han Dynasty recorded their observations of Halley's comet in the Book of Han during October and November of 12 BC.

The ancient Greeks studied everything, so it was not long before they gave comets and their sudden appearance an explanation. In

500 BC, the Greek philosopher Anaxagoras believed comets were really just faint clusters of stars. Much later, Aristotle was of the opinion that a comet was caused by the sudden ignition of dry pockets of air in the upper atmosphere. He reasoned that the Sun, Moon and planets always travel along the ecliptic in the night sky and so are confined to the Zodiacal constellations. By contrast, comets could appear anywhere and moved on paths well away from the ecliptic; this, he concluded, was because they were not a celestial phenomenon.

There was not universal agreement with this theory. The Roman philosopher Seneca (4 BC–AD 65) observed that when a comet was present in the night sky its motion was completely unaffected by the wind, suggesting it was a celestial phenomenon. In the end, it was Aristotle's argument which survived through history and, as late as the 16th century, Tycho Brahe was able to show that comets were nothing to do with the Earth's atmosphere, but external objects well beyond the orbit of our world.

In medieval times, superstition was a way of life. Virtually anything in the slightest way out of the ordinary was viewed with suspicion and mistrust. During the Middle Ages, comets were associated with bad fortune once more. In particular, they were thought to mark the deaths of kings and noblemen.

Halley's comet appears on the Bayeux Tapestry. The picture depicts some helpful souls telling King Harold II of the appearance of the comet in the skies of England shortly before the Battle of Hastings in 1066 – not the sort of thing you want to hear if you are of a superstitious disposition. As is well known, the battle was not a success for Harold, and the defeat no doubt reinforced the belief further that comets were to be regarded as a sign of bad luck.

By the 17th century, it was known that comets were external bodies, and the main concern (as it still is today) was that a comet could collide with the Earth and destroy all life here. This

Halley's comet, as featured on
the Bayeaux Tapestry.

sparked a number of 'end of the world' predictions. Interestingly, one came from the Lucasion Professor of Mathematics at Cambridge, William Whiston, who had succeeded Sir Isaac Newton. Whiston predicted the world would come to end at some point on 16 October 1736 when a comet collided with the Earth. Needless to say, that prediction can be added to the vast list of unrealised prophecies.

In 1773, the French astronomer and populariser Joseph Jérôme Lefrançois de Lalande wrote a technical paper about comets which seems to have been misinterpreted as a prediction of a comet collision, this time on either 20 or 21 May 1773.

The Great Comet of 1811 was the brightest comet for many years, and remained visible in the skies for a staggering 260 days. The comet was discovered by the French astronomer Honoré Flaugergues on 25 March 1811. It caused a stir in other circles: the English poet and artist John Linnell seems to have been inspired by his sighting of it and produced a drawing; indeed it may even be this comet which appears in his famous panel called 'The Ghost of the Flea'. The comet was also thought to be a bad omen by some, and its malign appearance was thought to have foretold the invasion of Russia by France. The Great Comet also provides an instance of a comet being associated with good luck. It seems that 1811 was an excellent year for wine-making, and wine merchants sold their 'comet wine' for a good many years afterwards at rather inflated prices.

The Great Comet of 1811 has an orbital period of some 3,757 years, so it will not be returning to Earth until the 56th century. Perhaps there will be some comet wine ready to welcome its next visit.

The 1910 return of Halley's comet saw the rise of something which, sadly, is still with us today: *comet hysteria*. By the time of the 1910 appearance, a number of new scientific tools and techniques had been developed. Photography had been sufficiently developed to allow the first photographs of the comet to be made. The powerful tool of spectroscopy had also been developed – the technique whereby the light of an object is split into a spectrum, and the fingerprints of the chemical elements present in the object are shown (see Chapter 4).

When the spectroscope was applied to Halley's comet, the poisonous agent Cyanogen was discovered to be present. Earth was due to pass through the tail of the comet, and a popular astronomer of the day, Cammile Flammarion, wildly speculated that vast amounts of Cyanogen would enter the Earth's atmosphere and, as a result, all life on Earth would perish. This brought all manner of crackpots and their inventions to the fore. Suddenly comet gas masks were on the market, along with 'comet pills' and even comet umbrellas (quite what good these would do was never made clear). A number of more level-headed astronomers pointed out that the amount of Cyanogen which would enter the atmosphere was incredibly small and, of course, the 1910 encounter with Halley passed without incident.

It would be nice to think that comet hysteria was confined to a time when few people understood astronomy. Today, facts and figures about comets can be obtained from a variety of sources; astronomy and science in general have become popular subjects, and yet the fundamental backbone of science – rational and objective thought – often seems to get lost in favour of gloss. Pseudoscience persists and has used the same media and science to continue its presence.

While many independent thinkers are harmless, there are others who are not; a tragic example is the case of the Heaven's Gate cult. This UFO religious cult was based in California and was founded by Marshall Applewhite and Bonnie Nettles at some point in the early 1970s. In 1997, the outstanding comet Hale-Bopp was a striking presence in the night skies, and Applewhite believed that an alien spaceship was following behind the comet. He and his 38 followers committed mass suicide together in the belief that they would be rescued by this alleged spacecraft. This affair is a sad reminder that, even in the modern age, comet hysteria and pseudoscience are still with us.

THE COMET SEEKERS

As telescopes improved, and it was realised that comets were external bodies out in the Solar System, so astronomers decided to look for new ones. Perhaps one of the greatest comet hunters of his time was Jean-Louis Pons, a French astronomer. Pons was an excellent observer and he went on to discover no fewer than 37 comets.

Another comet hunter was Caroline Herschel. Caroline was the sister of the great Sir William Herschel, who had discovered the planet Uranus. At the age of 10 she contracted Typhus which stunted her growth considerably. She moved from Hanover to join William at Bath, and became an invaluable assistant, helping him with his observations and mirror grinding. Caroline would become an important astronomer in her own right, and she went on to discover eight comets.

Another interesting comet hunter from this era was Charles Messier. Messier was a French astronomer, and discovered 13 comets, and yet it is not his comets that he is remembered for. Messier searched the sky for comets, but he frequently came across objects like galaxies and nebulae which appeared comet-like, but were nothing of the sort. He got so fed up with these objects that he compiled a list recording their positions so he and other astronomers wouldn't confuse them with comets in the future. This list is the famous Messier list which contains the details of the brightest and most spectacular deep-sky objects visible in the northern hemisphere.

Since the 19th century, many astronomers both amateur and professional have gone on to discover a vast number of comets. Comet hunting and monitoring continues to be one of the staples of modern amateur astronomy. It is a job which will surely never come to an end.

NOTABLE COMETS

There have been a good number of notable comets in recent times, and to record every instance would use up a great deal of space. The list below is in no way exhaustive; rather it gives a feel for some of the more dramatic past comet encounters.

Halley's comet. As we have seen, this is a bright periodic comet returning every 75 years or so. It was also the first comet to be identified as a periodic comet. Sir Edmund Halley examined the historical records concerning the appearances of comets. Halley calculated the orbit of the bright comet of 1682 and realised it had almost the same orbit as the one seen in 1531 and that of 1607. This led him to believe that all three instances were of the same comet, which had an orbital period of 75 years. Halley predicted that the comet would return in 1758, which it duly did.

Kohoutek's comet. This is a perfect example of a comet not doing as predicted. It was discovered by Luboš Kohoutek, a Czech astronomer, on 7 March 1973. Kohoutek was hyped up to be the 'Comet of the Century' by the media and some astronomers, and as a result public expectations were high. Alas the comet had other ideas and in the event was nowhere near as bright as expected and was regarded as a great disappointment.

Comet West. This comet was discovered by Richard West on 10 August 1975. By the time it came to perihelion on 25 February 1976, West's comet was rather bright, indeed on 26 and 27 February 1976, it was bright enough to be seen in the daytime skies. Unfortunately, thanks to the vast disappointment of Kohoutek's comet, comet West was largely ignored by the media.

Comet Hale-Bopp, as photographed by Andy Gannon.

Comet Hyakutake. This was the bright comet of 1996. Hyakutake made one of the closest cometary passes to Earth. It was discovered by Japanese amateur astronomer Yuji Hyakutake on 31 January 1996 using an enormous pair of binoculars. (Each objective lens had a diameter of 6 inches. The binoculars were mounted on a rigid mount, of course.) The comet was bright for a couple of days and, at its closest approach to Earth, its motion in the sky could be detected after a few minutes.

Comet Hale-Bopp. This is one of the great comets of the 20th century. It was discovered independently by Alan Hale and Thomas Bopp on 23 July 1995. The comet really was a stunning sight – easily visible with the naked eye, while a small pair of binoculars revealed the gas tail as well as the dust one. It was one of the most widely observed comets of the 20th century and was visible with the naked eye for a record 18 months.

Comet McNaught (C/2001 P1). Also dubbed the Great Comet of 2007, this comet was discovered by Robert McNaught on 7 August 2006. As it reached perihelion, the comet was so bright it could be seen in the daytime skies all over the world. McNaught was probably the brightest comet of the last 40 years. Photographs of it show its magnificently structured tail.

ICY VISITORS

Comets, like asteroids, are small bodies left over from the formation of the Solar System, with the distinction being that comets are made predominantly of water ice with only small amounts of rock. This composition immediately tells us something about their origin, specifically that to contain so much water they must have formed in the outer parts Solar System. It is thought that comets formed at distances ranging from the orbits of Jupiter and Saturn right out to the Kuiper Belt. Being so small, these icy bodies are easily affected by the gravity of the giant planets, and many would have been flung into different orbits, either plummeting in towards the inner Solar System or sent outwards.

Comets are very small, typically tens of kilometres across at the most, and are only easily observed when they come close to the Sun. As they get closer, sunlight causes solid material either on or just under their surface to heat up and turn into vapour. This spreads outwards, forming a coma that can stretch for thousands of miles, and the pressure of sunlight alone pushes the material away in a tail. This is why comets' tails always point away from the Sun, rather than trailing behind them in their orbits. The closer to the Sun, the more active they are, and the more spectacular they can become. One group, called the Kreutz

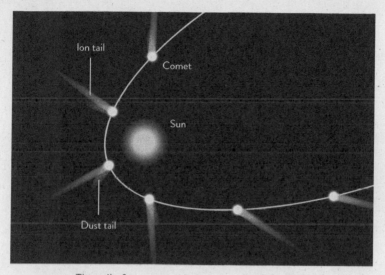

The tail of a comet as it moves around the Sun.

family comets, lie on orbits that take them incredibly close to the Sun, typically within a few thousand kilometres. Satellites such as SOHO, Stereo and SDO have observed a number of these passing in front of or behind the Sun, with a range of outcomes. Some survive more or less intact, though others either hit the surface of the Sun or are broken up as they pass. It is thought that these 'kamikaze comets' originate from one larger comet which broke up relatively recently, perhaps just a few hundred years ago, leaving many smaller objects on the same orbit.

It isn't always comets that get close to the Sun that can experience dramatic changes. The Sun's heat and light can on occasion release gas and dust from relatively large pockets beneath the icy surface, creating an outburst. A recent, and very memorable, event was a spectacular outburst of comet Holmes, a fairly faint comet just a few kilometres across which resides between the orbits of Mars and Jupiter. In October 2007, it suddenly brightened by a factor of about half a million, evidence that a huge amount of material had been released from the nucleus. The cloud of newly released material was easily observable in binoculars or a small telescope for several months, though as it spread out it became fainter and harder to spot. Astonishingly this expanding shell grew until it was well over a million kilometres in size, briefly making comet Holmes the largest object in the Solar System – bigger even than the Sun. This was not the first time comet Holmes had done this, and in fact it was discovered as it experienced an outburst in 1892. It seems that some comets are more susceptible to outbursts than others, suggesting that it is due to some aspect of their structure.

Comets are split into a number of groups, or families. Short period comets have orbits of less than 200 years, and generally don't stray much further than the outer planets. It is thought that these comets were formed out in the Kuiper Belt, beyond the orbit of Neptune. Most short period comets orbit in the same plane as the planets, but some of them have been caused to head into the inner Solar System by the gravitational influence of the giant planets. In particular, some are on orbits that take them out to the distance of Jupiter's orbit, and are called Jupiter family comets. These comets tend to have orbits lasting between five and ten years.

Long period comets have periods longer than two hundred years, and many have orbits which are *much* longer. Long period comets come in from all directions, and aren't associated with the orbits of any planet. They are thought to come in from a distant region known as the Oort cloud. While it has never been directly observed, the range of directions from which long period comets originate means that the Oort cloud is predicted to have a roughly spherical structure. This is to be expected of a group of bodies which formed near the orbits of the giant planets, particularly Jupiter and Saturn, and were ejected by interactions with them. It is not known what causes Oort cloud comets to venture into the inner Solar System, but the favoured theory is that the passing stars scatter them inwards. In fact, some have predicted orbits which are so large that these comets could be travelling so fast that they even manage to escape the Solar System, though this has never been definitively confirmed. Since such comets can take millions of years to get close to Earth, identifying specific events in their past is essentially impossible. It is thought that some long period comets have previously been affected by giant planets and placed on shorter orbits, essentially becoming short period comets. Such comets, which include Halley's comet, still have highly inclined orbits, and this is how they are distinguished from the proposed Kuiper Belt comets.

More recently, in the last few decades, astronomers have discovered 'main belt comets', which are on roughly circular orbits in the asteroid belt. Distinguishing such objects from normal asteroids is relatively hard, and they have tended to be identified by the presence of a tail. While some are the results of asteroids colliding, others do seem to be made of ice. Finding such icy bodies so close to the Sun was unexpected since the ice should have been baked away long ago, and there are many questions as to how they have survived so long. It is slowly becoming apparent that comets can form in all sorts of places, and that perhaps they have a great deal to tell us yet about their history as well as that of the Solar System. The difficulty is that the various theories about the origin of comets are all rather hard to prove, and it is only close-up and in-depth studies of comets which are starting to answer this question.

IMPACT EARTH

While it is possible for comets to hit the Earth, the fact that most of them reside at such great distances means that they are generally considered much less of a danger than asteroids. But in the early

Solar System, when millions of comets were being flung around by interactions with Jupiter and Saturn, it is certain that Earth would have been a target. It is even possible that such collisions are responsible for the water we see on our planet.

We know that the early Earth had a hot, molten surface, and that most of the water would have disappeared into space. Once it cooled, the impact of icy bodies, such as comets, would have delivered water and created much of the Earth's oceans and atmosphere. A key test of this theory is whether the water in comets has the same precise composition as the water on the Earth. One way of determining this is to measure how much deuterium – a form of 'heavy hydrogen' – is in the water. Most of the comets studied in this way have not matched the composition of the Earth's water, containing more deuterium than our oceans, casting doubt on the theory. But in 2010 the Herschel Space Observatory studied the composition of comet Hartley 2 in detail and showed that its water matches ours. It is thought that Hartley 2 may have formed out in the Kuiper Belt, whereas the comets studied previously are Oort cloud comets, having formed closer in to the Sun. One possibility is that it is these Kuiper Belt comets which primarily delivered water onto Earth, rather than Oort cloud comets, though why this would be is not clear. Another possibility is that comets are not the dominant source of the Earth's water at all, and that most of it came from asteroids. We are a long way off proving either theory, since there have only been just over a handful of comets studied in this way, but that doesn't stop lots of speculation. There has long been the idea that life may have formed on the surface of comets which subsequently impacted on the Earth and 'seeded' it, though again this theory is very hard to prove.

One of the side effects of the tail of a comet is that they leave a trail of material around their orbit, slowly spreading out to fill a broader trail around the Sun. This material is in the form of tiny grains of ice and dust, and if the Earth passes through this path of material we see a meteor shower. It is quite astonishing that such a tiny piece of space dust, often no larger than a grain of sand, can create the bright flash of a shooting star, more correctly called a meteor. The reason is that the Earth passes through these cometary trails at tens of thousands of miles per hour. As the particles enter the atmosphere at these supersonic speeds they compress the air in front of them to searingly high temperatures. This compressed and heated air glows to create the flash of light, but also ablates the surface of the incoming object, vaporising its surface. In most cases, when the particles are small enough, they are completely destroyed and never reach the ground. All that is normally seen is a brief streak across the sky, normally lasting for no more than a few seconds, after which there is no visible mark. Now and again there are much brighter flashes caused by larger objects, and these can leave a faint ghostly train which hangs in the air for some time. These are referred to as fireballs and are caused not by cometary material but by small asteroids, more like the size of a brick or a football. These tend to travel more slowly and persist for longer, and as a result are often observed by relatively large numbers of people over wide areas.

It is important to point out that not all meteors originate in the trails left by comets, as there are many other pieces of dust and rock out there as well. The flashes created by this other material are called 'sporadic meteors', as they appear to come from all directions. On any given night, it is not unusual to see several of these per hour. Meteors associated with a comet, however, all come from the same direction, and so share a common point of origin. As the Earth moves through the cometary debris field, the meteors all come through the atmosphere from the Earth's direction of travel. At any given time of year this is a particular spot in the sky, and meteors in a shower appear to radiate from that point, called the 'radiant'. The names of meteor showers are referred to by where this radiant is. In mid-August, for example, the Earth passes through the trail of debris left by comet Swift-Tuttle. At that time of year, the Earth's direction of travel around the Sun is towards the constellation of Perseus, and so this shower is known as the Perseid meteor shower. Similarly, in mid-November we pass through the dusty remains of comet Temple-Tuttle, and since at that point we are moving towards the constellation of Leo these are called the Leonids.

There are many showers throughout the year, associated with different comets. Many have distinctive properties, such as particularly bright or slower-moving meteors, normally related to the cloud of material as well as the way in which the Earth is passing through it. The cloud of material can be thought of as a tube of dust hanging in space, and the shower lasts for as long as the Earth is passing through this tube. Parts of the tube, often near its core, are denser than others, and these are when we see the peak of meteor showers, which can last anywhere from a few hours to several days. Observations of the numbers, directions and brightness of meteors during a shower can be used to trace the density of the clouds of material, and be used to predict how future showers may vary.

VISITING COMETS

While apparitions of comets and meteor showers can tell us a lot, there is much we can't see from here on Earth. For a start, all we ever really see is the bright coma, a cloud of ice and dust that is much, much larger than the solid nucleus of the comet. To learn more, we have to travel to the comets, and

Jets of material spewing from the surface of Halley's comet,
seen by the Giotto spacecraft.

this has been done by a number of spacecraft over the years.
The first visited comet was the most famous: Halley's comet.
As it made one of its regular visits to the inner Solar System
in 1986, Halley was visited by NASA's twin Vega spacecraft
(continuing their missions past Venus) and ESA's Giotto probe.
The images provided the first ever close-up view of a comet's
nucleus, revealing an irregular, peanut-shaped object around 15
km long. Surprisingly, the surface of the nucleus was not bright,
as one might expect a largely icy body to be, but very dark – in
fact darker than coal. And rather than material being vented
from all over the surface, only around a tenth of it seemed to
be active, with seven individual jets of material being identified.

The dark surface suggested a thick coating of dust, and observations showed that it is very rich in carbon. Studies of the comet's composition showed that only around 80 per cent of it is made of water. Most of the rest of the material is carbon monoxide (a molecule composed of carbon and oxygen), with small amounts of methane (carbon and hydrogen) and ammonia (hydrogen and nitrogen). In general, the make-up of the comet is very similar to that which makes up the Sun, confirming that it really is pristine material from early in the Solar System's history.

A total of five comets have been visited by spacecraft, including Halley. In 2001, the Deep Space 1 mission passed by comet Borrelly, showing an irregular but less active body, while Stardust flew through the tail of comet Wild 2 in 2004. Stardust collected tiny particles of cometary dust in a sponge-like material called aerogel, and returned them to Earth in 2006. A laboratory study of the tiny dust grains showed that the makeup of comets is much more complex than previously thought. While they are thought to have formed from ice and dust in the outer Solar System, the properties of the grains show that they formed in a much wider range of conditions. The cometary particles were in general much larger than expected, with some of the larger crystalline grains showing signs that they formed at temperatures of over a thousand degrees. These must have originated much closer to the Sun than the icy particles that formed at much lower temperatures, and been pushed outwards by the pressure of the Sun's light. Surprisingly complex molecules were also found, including the amino acid glycine, one of a number of key building blocks of life here on Earth.

In 2005, the Deep Impact probe didn't just visit a comet, but actually took a pot-shot. As it passed comet Tempel 1, it released a 370-kg impactor, made primarily of copper, which hit the nucleus. The impact caused a huge amount of ice and dust to be released, and the composition of the material showed that Tempel 1 probably formed at the same distance from the Sun as Uranus and Neptune before being sent in towards the inner Solar System where it now orbits. Unfortunately, the huge amount of material released by the impact meant that the Deep Impact spacecraft itself was unable to image the crater before it departed. Telescopes on Earth and in space continued to observe Tempel 1 and continued tracking the venting of material for weeks after the impact, with the total amount of material released estimated to exceed 250,000 tonnes.

As a result of Deep Impact's poor view of the crater it excavated on Tempel 1, NASA directed its Stardust probe to swing by and take a better look. It passed by in 2011, around six years after Deep Impact's visit, and showed a tiny crater around 45 metres across – barely more than a speck in Stardust's images. It observed parts of the surface not seen by Deep Impact, and even showed subtle changes that had occurred over the course of the comet's orbit around the Sun.

The most recent comet to have been visited is comet Hartley 2, which passed the Earth in November 2010 on its regular journey through the inner Solar System. In another example of interplanetary recycling, the Deep Impact spacecraft passed by and revealed perhaps the most peculiar comet to date. Hartley 2 seems to be made of two separate objects, joined in the middle by a smooth collar of material. The two 'lobes' are where the active jets originate, and the composition of the material indicated that they were indeed originally two distinct bodies.

The most ambitious mission to a comet is still to come. In 2004, ESA launched its Rosetta spacecraft, and in 2014, after flybys of Earth, Mars and two asteroids, it will encounter comet Churyumov-Gerasimenko. Unlike any previous mission, Rosetta will not simply fly past but will enter orbit around the comet – becoming the first spacecraft to do so. By remaining in orbit for over 18 months, Rosetta will monitor how the comet's nucleus and coma evolve as

The dumbbell-shaped comet Hartley 2, visited by
the Deep Impact spacecraft.

it makes its closest approach to the Sun in late 2015. An additional feature of the mission is a lander, named Philae, which will land on the comet and drill down 20 cm, allowing the study of material from beneath the surface. Such an in-depth study should shed some light on where comets formed, or even whether they all formed in the same place.

* 14 *

One Among Many

For thousands of years, the only group of planets we knew existed were those in our own Solar System. Within the last twenty years, that situation has changed dramatically, with hundreds of planets known to exist around other stars. But speculation about other solar systems is far from a new idea. Mankind has debated the nature of the stars since we first noticed them glowing in the skies above, and philosophers have argued for centuries that those twinkling lights are similar in nature to our own Sun, but much, much more distant. As our understanding of physics has progressed over the years, astronomers have been able to fully comprehend the vast scales of the cosmos. We now know that not only is the Sun one among many, but the number of stars is far greater than the several thousand visible to the naked eye. It is estimated that there are a hundred billion stars in our Galaxy, and that there are a similar number of galaxies in the visible Universe. Not only is the number of stars unimaginably vast, far larger than the number of grains of sand on the Earth, but it may in fact be infinite.

Many of the philosophers who theorised that the stars are like our own Sun also postulated that they must therefore have their own family of planets. A few hundred years ago the thought of life on the Moon, Mars or even the Sun, was not as far-fetched as it is now, and this speculation extended to the innumerable worlds around other stars. As we have seen throughout this book, the planets in our Solar System are an eclectic mix of worlds, with a wide range of sizes, temperatures, weather systems and surface conditions. The observed variety led astronomers to wonder what worlds around other stars might be like. With the ability to bear life rare in our Solar System, and possibly confined to one planet, the prospects of life elsewhere might seem slim. But the sheer number of planets discovered, and the wide range of conditions presumed to be on their surfaces, means that, even if life is very rare, there could still be millions of alien species out there.

SEARCHING FOR OTHER PLANETS

Searching for planets in other Solar Systems is incredibly difficult, since the planets are dwarfed by their stars in terms of mass, size, brightness, and pretty much every other way in which they might be measured. Until recently extrasolar planets had never been seen directly, and the images to date are only achieved by blocking out the light from the star itself. What's astonishing is that there are almost a thousand confirmed examples of planets around other stars – and the number is growing so rapidly that by the time you are reading this it may already have leapt well beyond that particular milestone. But if they are so small and faint, how have they been detected at all?

Most of the planets discovered to date have been found by one of two methods. The first is the 'transit' method, which involves looking for planets which pass directly in front of their star as viewed from here on Earth. As they do so, they block out a tiny bit of the light – in exactly the same way as Venus and Mercury do when they pass in front of the Sun. But being so distant we can't see a small black disc, and all we see is a slight dimming of the light from the star. It is much easier to find larger planets by this method, since they block out a larger fraction of their parent star's light. While most planetary systems are not oriented in such a way that their planets pass in front of their star, this method has allowed the detection of hundreds of planets. Although most are much larger than the Earth, missions such as the Kepler satellite have been able to spot the passage of planets smaller than the Earth pass in front of their star.

A more successful method is to look for the wobble of the star as the planet moves around it. As with all orbiting systems, the planet does not technically orbit the star, but instead both the planet and star orbit their combined centre of mass. Since the planet is much lighter, it moves in a much wider orbit, but the star does move a little. In our Solar System the motion of Jupiter, for example, actually causes the Sun to move around a point near its surface, rather than remaining perfectly stationary. The effect is so small that it is incredibly rare to actually see the star 'wobbling' in the image, but instead something slightly more subtle is searched for. As the star moves in a small circle it spends some time moving towards us and some time moving in the opposite direction. This motion causes the light from the star to be shifted in wavelength, or colour, by tiny amounts through an effect called the Doppler effect. The change in wavelength is incredibly small, though it is more significant for more massive planets, and also for those orbiting more closely to the star.

There is a third method used to find planets, which relies of the process of gravitational lensing. This is where the passage of a planet in front of a much more distant, unrelated star causes the light from this background to be warped and magnified. Such chance alignments happen from time to time and have been used to detect planets around stars thousands of light years away, but since it is very hard to predict it has not led to anywhere near the number of discovered planets as the two previous methods.

The various detection methods are often combined to learn more about a planet. The transit of a planet in front of its star tells us about its size, while the amount of wobble it generates in the star tells us how massive it is. Together, these two pieces of information allow the density of the planet to be calculated. Specifically, we can establish whether it is a gas giant, like Jupiter or Saturn, or a rocky planet like the Earth and Mars. The time it takes the planet to complete an orbit can be used to determine the distance from its Sun, and so the temperature on the surface can be estimated. But there are many factors at play in the surface conditions, not least of which is the presence of an atmosphere. Future missions will be aimed at characterising the composition and conditions on the surfaces of extrasolar planets by studying their light in great detail. Being so close to bright stars, this is very hard to do, but by no means unachievable, particularly for planets around fainter stars such as red dwarfs.

Many of the planets discovered around other stars are not alone, but orbit in groups, making up planetary systems. One of the most surprising things about these remote groups of planets is that very few have been found to resemble our own Solar System. Many have massive planets orbiting very close to their stars, probably having migrated inwards during the formation of the system. Others have a number of rocky planets orbiting so close to their stars that their surfaces must be baked dry. We've even detected planets around pairs of stars, called binary stars. This includes planets orbiting a pair of very close stars, as well as planets orbiting one star with a second star much further away. One such system even involves four stars. How and why these planetary systems form is a matter of intense study, and should eventually shed some light on the processes that led to the evolution of our own Solar System into the state we see it in today.

LIFE ON EARTH

Microbial life first appeared on Earth within the first billion years of the planet's existence – almost as soon as the conditions allowed. It's not known exactly how this happened or what the spark of life was – such a discussion is well beyond the scope of this book – but there seem to be a few key requirements for life to exist. The first, and seemingly most important, is liquid water, which seems to be essential for complex chemical reactions. First of all, its structure allows a huge range of different molecules to dissolve in it and mix together. Second, the hydrogen and oxygen that make up water are key constituents of DNA which makes up all life on Earth, the others being carbon, nitrogen and phosphorus. To form complex life, such as animals and plants, it is important to be able not only to generate energy but also to store it for use later – just like a rechargeable battery. Most life on Earth stores energy in sugars, or carbohydrates, which are a combination of carbon, hydrogen and oxygen.

There are numerous ways of creating carbohydrates, one of the most simple of which is the photosynthesis used by plants. Light from the Sun is used to split water up into hydrogen and oxygen. This is combined with carbon dioxide to produce carbohydrates, which can be used as an energy source by the planet, and oxygen, which is returned to the atmosphere. It is this process, combined with the dissolving of carbon dioxide in the oceans, which has made the Earth's atmosphere so perfect for the evolution of animals, leading (eventually) to us.

Life is obviously more than just water, and a wide range of other chemical elements are needed for complex organisms to form. A very useful source of these other chemical elements is a family of compounds that include carbon, hydrogen and oxygen. Not only are they crucial to life, but they are often all that is left after billions of years of decay – they form the main constituents of coal and oil. Their importance to life is why these are referred to as 'organic chemicals'.

Finally, life requires a source of energy. Most life forms on Earth use sunlight, but there are some that use the heat from within the Earth – geothermal energy – to drive their chemical reactions. This source of energy is the third of the three requirements that we think are required for life to form, but the exact process by which this happened is still somewhat of a mystery. Experiments in the lab have taken water and organic chemicals and provided a source of energy, but failed to create life. There seems to be some spark missing, and experiments have been under way for decades to try to create artificial life, though so far with no success. Some people have theorised that a spark is exactly what is needed, and that bolts of lightning kick-started the first chemical reactions – though the true cause is still unknown.

Regardless of the process involved, it is these three basic requirements that are currently the focus of our search for life elsewhere. Both within our Solar System and beyond we are searching for places where there exists liquid water, organic chemicals and a source of energy. So far that search has been in vain. Such measurements are very hard to make, and even more so from a remote distance. We are only now nearing the capability to design robotic spacecraft and landers advanced enough to search for life within our Solar System, and travelling to another star system is impossible – at least at present. One thing is certain: if life does exist elsewhere in the Solar System it is in a very primitive form, probably more like the microbial life that formed on Earth almost

4 billion years ago. In the long run, however, it's hard to believe that the search will remain in vain. With so many hundreds, and possibly thousands, of planets discovered in other Solar Systems, there are bound to be many which have conditions similar to those on early Earth.

IS THE EARTH UNIQUE?

The Earth is certainly unique in our Solar System, but we are now able to detect the presence of small, rocky planets orbiting other stars. The fast pace of discoveries means that quoting record-breaking planets (such as the smallest found so far) will ensure that this book is immediately out of date. But what is not changing is that much of the effort is focused on finding planets like the Earth. These planets must be small enough to be rocky (rather than giant planets like Neptune and Jupiter), and be at the right distance from the Sun to allow liquid water to exist on their surface. To this end, the searches often focus on the 'habitable zone' around the parent star, where the ambient temperature is not too dissimilar from those on Earth. For smaller, fainter stars this habitable zone is closer in to the star, while for larger, hotter stars it is further out. With the current techniques, it is easier to find such planets around smaller, less massive stars, as their smaller orbits means they have a more easily detectable effect on the star.

But being in the right place might not be enough, as the atmosphere of the planet has a critical impact on the surface conditions. Take the Earth, for example. Without its atmosphere, the surface would be around 20 degrees cooler, and would therefore probably be covered in ice rather than liquid water. The atmosphere means that the average temperature on Earth is 10–20 degrees above zero, but if it were much thicker, or contained more greenhouse gases such as carbon dioxide and methane, the surface might be more like that of Venus: searingly hot and completely dry. Our current ability to study the atmospheres of planets around other stars is very limited, though there are missions planned to do just that. These missions, if they happen, will give us a much better idea of the conditions present on the surfaces of these planets, and whether liquid water and organic chemicals are likely to exist.

Beyond the three basic requirements, water, organic chemicals and an energy source, there could be other more subtle effects which have made Earth an ideal home, but which we have not fully appreciated the importance of. A key example of this might be the planet's magnetic field, which is generated by the motion of molten iron in the core and protects the atmosphere and surface from harmful radiation from space. Without this magnetic field, high energy particles originating from the Sun and elsewhere in the Galaxy can break apart the molecules and strip atoms of their electrons. Microbial life is relatively insensitive to this, but more complex life can suffer serious effects. We also know that this magnetic field isn't a permanent feature, and has varied over time. Studies of the geological record show that the Earth's magnetic field has flipped many times in the past, with the magnetic poles switching direction and pointing the other way. It is not caused by any outward effects, but rather by changes in the circulation of the molten material within the Earth's core. This magnetic flipping has occurred thousands of times in the Earth's history, and the fact that complex life has suffered no serious consequences indicates that it is not a simple case of the field 'switching off' and returning in the opposite orientation. More likely, the magnetic field becomes much more complex for a while, with several 'poles' distributed over the Earth, before settling down into the opposite arrangement. There seem to be no severe effects on life, complex or otherwise, though the navigation systems of pigeons might be adversely affected.

NATURE VS. NURTURE

As well as the evolution of the planet itself, there are also external factors that might make the Earth a more unusual place than it first appears. For example, it may also be that the Sun itself is somewhat special. Studies have shown that our star seems to be slightly more stable than other stars of a similar mass. Perhaps this means that it emits less radiation than most other stars, making the Solar System slightly more hospitable. Somewhat ironically, much of this evidence comes from the searches for other planets, which need to take into account the variability of the host stars to pick out the weak signature of orbiting bodies. Closer to home, the presence of our relatively large Moon has meant that the Earth's rotation has stayed relatively stable, with the planet spinning about pretty much the same axis for billions of years. The conditions in a given region of the planet have (historically, at least) changed relatively slowly as the climate has varied over time, giving life time to adapt at a leisurely place. Other planets, such as Mars and possibly Venus, have changed their orientation much more dramatically than the Earth, causing the conditions on their surface to experience huge shifts. Perhaps these shifts have been too fast for life to evolve before finding its environment becoming inhospitable. Further out in the Solar System, we have the planet Jupiter, which uses its huge mass to sweep up many of the comets that come in from the outer Solar System. Without this planetary sentry, the Earth might have been bombarded with cometary debris for a much longer period of its history, wiping out advanced life in global extinction events.

These ideas are all fairly speculative at the moment, since without a large number of Earth-like planets to study in detail, we are dealing with a single case study: Earth itself. Perhaps these effects are important, and the Earth just happened to be the best of all worlds. Or maybe they have only a minor effect, and life really is able to take hold and evolve almost anywhere. It is hard to know without being able to study a great many other Solar Systems in detail. This doesn't only mean finding a planet like the Earth, with similar surface conditions, but also understanding the role that the rest of the planetary system could have had on the evolution of the planet over eons of time.

The origin of life is certainly a matter of hot debate, and often delves into the realm of philosophy and theology. Scientifically speaking, there are two prevailing opinions among astrobiologists: one is that life can form almost anywhere, and will be very common throughout the Universe, while the other is that the Earth just got lucky, and the chances of life forming are so slim that we are likely to be alone in the Universe. Most, however, are holding out on placing any bets until we've been able to search more thoroughly.

Glossary of Useful Terms

Aphelion: The position in the orbit of a planet or other body which is furthest from the Sun.

Asteroid: A small Solar System body. Most of the best known asteroids move around the Sun between the orbits of Mars and Jupiter.

Aurora: or 'polar lights', specifically Aurorae Borealis (northern lights) and Aurora Australis (southern lights). They occur in the Earth's upper atmosphere, and are caused by charged particles emitted by the Sun.

Axial tilt: The tilt of a planet's axis of rotation relative to its orbit.

Chromatic aberration: A false colour halo seen around objects in some refracting telescopes, particularly early ones, due to the different colours being refracted by different amounts. Modern refractors often have two or three lenses cemented together and, as a result, chromatic aberration can be almost (though not entirely) eliminated.

Comet: A small icy body in orbit around the Sun, left over from the creation of the planets. When comets come close to the Sun they can lose material, creating the characteristic tail.

Constellation: A pattern of stars in the sky. The stars in any constellation are not really related to each other, because they are at very different distances from the Earth.

Corona: The outermost part of the Sun's atmosphere, made up of very tenuous gas. It is visible with the naked eye only during a total solar eclipse.

Coronal mass ejection (CME): The release of large amounts of material from the surface of the Sun.

Cryovolcanism: A process similar to volcanism on Earth, but which erupts icy materials rather than molten rock.

Day, solar: The mean interval between successive meridian passages of the Sun, i.e. the points at which the Sun is due south. The length of a solar day varies through the years, but on average is it 24 hours long. It is longer than the sidereal day because the Sun seems to move eastward against the stars at an average rate of approximately one degree per day. This is because as the Earth spins it also moves around its orbit, and must rotate a small amount extra every day for the Sun to appear in the same place.

Day, sidereal: The interval between successive meridian passages, or culminations, of the same star: 23 hours 56 minutes 4.091 seconds. This is the true rotation rate of the Earth.

Dwarf planet: An object orbiting the Sun that has a sufficiently high mass to have become spherical, but which has not cleared its own orbit of other material.

Eclipse, lunar: The passage of the Moon through the shadow cast by the Earth. Lunar eclipses may be either total or partial. At some eclipses, totality may last for approximately 1¾ hours, though most are shorter.

Eclipse, solar: The blotting-out of the Sun by the Moon, so that the Moon is then directly between the Earth and the Sun. Total eclipses can last for over 7 minutes under exceptionally favourable circumstances. In a partial eclipse, the Sun is incompletely covered. In an annular eclipse, exact alignment occurs when the Moon is in the far part of its orbit, and so appears smaller than the Sun; a ring of sunlight is left showing round the dark body of the Moon. Strictly speaking, a solar 'eclipse' is the occultation of the Sun by the Moon.

Ecliptic: The apparent yearly path of the Sun among the stars. It is more accurately defined as the projection of the Earth's orbit on to the celestial sphere. The planets move through the sky close to the Ecliptic plane.

Elliptical orbit: The orbits of all planets and moons in the Solar System are not circular but elliptical, meaning they are slightly squashed in shape. The amount the circle is squashed by is called the eccentricity, which can take any values between 0 and 1. An ellipse with no eccentricity (i.e. $e = 0$) is a circle.

Exoplanet, extrasolar planet: A planet orbiting a star other than the Sun.

Inclination: In the context of orbits, the tilt of the orbit relative to some reference plane. In the Solar System this is relative to the Earth's orbit.

Filter: In astronomy, a device that only lets through specific wavelengths of light. Filters can be 'broadband', e.g. letting through red light, or 'narrowband', letting through only a narrow range of wavelengths, usually a range emitted by a particular element.

Finderscope: A small, wide-field telescope attached to a larger one, used for sighting purposes.

Fraunhofer lines: The dark absorption lines in the spectrum of the Sun, or any other star. The lines are caused by the absorption of light by particular elements and can be used to deduce a star's composition.

Focal length: This is the distance from a mirror or lens to the point where light rays are focused.

Kuiper Belt: A belt of small bodies moving round the Sun from the orbit of Neptune and beyond.

Light year: The distance travelled by light in one year, which is around 10 million million km (6 million million miles).

Meteor: A piece of debris, usually from a comet. When it dashes into the Earth's atmosphere, it burns away to produce a shooting star.

Meteorite: A solid body from space landing on the Earth. Most meteorites come from the Asteroid Belt.

Oort cloud: An assumed spherical shell of comets surrounding the Sun, at a range of about one light year.

Orbital period: The period that a planet or other object takes to orbit the Sun, or that a satellite takes to orbit its primary planet.

Perihelion: The position in the orbit of a planet or other body which is closest to the sun.

Photosphere: The visible surface of the Sun or other star.

Planet: An object orbiting the Sun or another star that has a sufficiently high mass to have become spherical and to have cleared its own orbit of other material.

Plasma: Ionised gas, normally at very high temperature. The surface of the Sun is made of plasma.

Precession: The movement of the axis of rotation of a spinning body such as a planet or moon, or of the orientation of its orbit around a parent body.

Resonance: A natural period of rotation or orbit of a body. In the Solar System, a resonance is usually in relation to a second object, such as another moon or planet, with one body's orbital period being a simple multiple of the other's.

Satellite: A natural or artificial object that is orbiting another astronomical body such as a planet.

Solar flare: A massive release of energy from the Sun's surface, often accompanied by a coronal mass ejection.

Spectrum: The range of wavelengths, or colours, of light that an object emits. The spectrum of an object can be used to deduce its composition.

Sunspot: A dark area on the surface of the Sun that is slightly cooler than its surroundings.

Telescope, aerial: These are very long focal length refractors and were used in the latter parts of the 19th century. They were built with very long focal lengths in an attempt to reduce chromatic aberration. Such telescopes were often very unwieldy and were abandoned as better optics were made.

Telescope, Newtonian reflector: This telescope was invented by Sir Isaac Newton. It consists of a tube, at one end of which is a concave mirror (the primary mirror), while at the other end is a flat secondary mirror. Light from a distant object comes into the tube and hits the primary mirror where it is magnified. The light then travels up to the secondary mirror where it is reflected further into an eyepiece. Here it can be brought into focus and magnified further.

Telescope, refractor: This is the simplest type of telescope. It consists of a tube, at one end of which is a large lens (the primary) and a smaller lens (the eyepiece) at the other end. Light from a distant object is magnified by the primary, and brought into focus and magnified further by the eyepiece.

Transit: The passage of one object in front of a larger one, blocking some or all of the light.

Trojan asteroid: An asteroid which shares a planet's orbit around the Sun, but is ahead or behind in the orbit by 60 degrees.

Index

Note: page numbers in **bold** refer to diagrams and photographs.

achondrites 260
Adams, John Couch 237–8, 252
Airy, George 237
Aldrin, Buzz 129
aliens 280
Almagest (Ptolemy) 35–6, 41
Amalthea 195–6
ammonia 240, 250, 292
Anaximander 33
ancient stargazers 23–36, 82, 96, 152, 178, 202, 276–80
angular momentum 110, 138–9
Anthe 215
Antikythera mechanism 144
Antoniadi, Eugène 86, 155
antumbra 140
Apollo missions 20, 75–6, 129, 139, 146–50, 260
Apophis 266
Ariel 245
Aristarchus of Samos 34
Aristotle 38, 277
Armstrong, Neil 129
asteroid belt 20, 253, 257, 272, 287
asteroids 253–8, 259–65, 272, 288
    collisions 261–2, 265, 289
    moons 261
    near-Earth 265–7
astronauts 20–1, 75–6
aurorae 69–70
Aztecs 96

Babylonians 28, 28, 82, 96, 152, 178, 202, 276
BepiColombo satellite 94
Bianchini, Francesco 99
binoculars 62, 64, 128, 130, 239, 283
Bode's Law 255
Borrelly (comet) 292
Brahe, Tycho 21, 45–8, 50–4, 276, 277
Bruno, Giordana 42–4, 59
Bunsen, Robert 79–80

Callisto 184–5, 185, 191, 194–5, 198, 200
Calypso 222
carbon 299–300
carbon dioxide 95–6, 112, 122, 163–4, 175, 299
Carrington, Richard 68–9, 76
Cassini, Giovanni 153, 179, 186–7, 204, 223
Cassini (spacecraft) 196, 201, 212, 214–15, 214, 219–20, 222–4, 226–32, 227
Cassini Division 208, 209, 210, 212–13, 217–18
Catholicism 42–3, 50
Centaurs 257
Ceres 253, 255, 263–5, 271

Challis, James 237–8
Chang'e 2 263
Charon 269, 272
China 25, **26**, 27, 60, 82, 149–50,
    152, 178, 202, 263, 276
chondrites 260, 263
Christianity 41–4, 50, 57–8
chromatic aberration 83
Churyumov-Gerasimenko 293–4
Cluster mission 72
comets 20, 46–7, 178, 197–8, **199**,
    275–94
    'comet hysteria' 279–80
    composition 284–7, 291–3
    impacts 275, 277–80, 287–90,
        303
    Jupiter family 286
    long period 286–7
    'main belt' 287
    missions to 290–4, **291**
    notable 282–4
    search for 281
    short period 286, 287
    tails 284–5, **285**, 288–9
constellations 24–5
Copernicus, Nicolaus 21, 38–45, 47,
    52, 58, 185
coral growth 134–5
cosmology 33–6
Crabtree, William 97–8
cryovolcanism 229, 250–1, 272
Curiosity (rover) 169–71, **171**,
    174–6
Cyanogen 279–80

Dactyl 261, **262**
Daphnis 215, 225
Dawn mission 263–5, **264**, 267
Deep Impact probe 292–3
Deep Space I mission 292
deferents 35, **36**
Deimos 158
Deneb 24, 31
deoxyribonucleic acid (DNA) 299
deuterium 288
dinosaurs, extinction 259–60

Dione 204, 210, 219, 220–2
Doppler effect 297
dwarf planets 20, 253, 268–73

Earth 17–18, 119, 120–2
    age 121
    and asteroids 257, 259–60, 265–7
    atmosphere 17, 120, 121–2, 301–2
    aurorae 69–70
    axis tilt 174, 303
    circumference 33
    climate change 77
    core 115, 121, 145
    crust 115, 120
    curved 33
    eclipses 139–43, **141**, **143**, 186
    formation 121
    and the formation of the Moon
        145–7
    and the geocentric model 34–7, **36**
    gravity on 133–4
    and the heliocentric model 40–3,
        47
    impacts 145–7, 265–7, 275,
        277–80, 287–90
    life on 120, 276, 288, 292,
        299–304
    magnetic field 117, 302
    mantle 115–16, 120
    and Mars 156
    and Mercury 87–9, **87**
    and meteorites 258–61, **259**
    orbit **16**, 52, 53, 120, 138–9
    Poles 120
    rotation 120, 138, 139, 303
    seasons 120
    and the Sun 69–70, 75–7
    surface 114–16, 120–2
    tectonic activity 115, 120–2
    temperatures on 17, 120
    tides 111, 133–9, **136**
    uniqueness 301–4
    volcanoes 18, 115–16, 121–2
    water on 116, 120, 276, 288,
        299–300
    weather on 76–7

Earthshine 127
eclipses 139–44, 141, 143
    annular 140, 141
    of Io 186–9, 188
    lunar 142–3, 143, 186
    partial 140–1, 142–3
    solar 64, 66, 140, 141, 144
    total 140, 141, 142–4
ecliptic 183–4, 277
ecliptic plane 141
Egypt, Ancient 25, 27–8
Einstein, Albert 82, 90, 142
Enceladus 201, 204, 210, 215,
    218–20, 221, 222, 229, 232
    fountains of 219–20
    life on 220
energy 299–300, 302
epicycles 35, 36
Epithemeus 225
equinoxes, precession of the 31, 36
Eratosthenes of Cyrene 32–3
Eris 271
433 Eros 262
ESA 72, 75, 94, 111, 116, 149, 176,
    200, 291, 293
Euclid 38
Eudoxus of Cnidus 34–5
Europa 184–5, 185, 191–4, 195,
    196, 200, 229
    life on 193–4
European Southern Observatory 108
ExoMars 176
extinction events 259–60, 279–80,
    303
eye, naked 64

Fraunhofer lines 78–9

Galatea 252
galaxies 295
Galileo Galilei 21, 56–8, 83–4, 97,
    153, 178–9, 186–7, 203
Galileo spacecraft 190–1, 193–4,
    195, 196, 198, 261, 262
Galle, Johann 238, 252

Ganymede 19, 178, 184–5, 185,
    191–2, 194–5, 198, 200
gas giants 19, 177, 253, 298
    see also Jupiter; Saturn
951 Gaspra 261
general relativity theory 82, 90, 142
geocentrism 34–6, 36, 37
geometry 29, 34, 38, 49
Giotto probe 291, 291
God 36, 49, 276
gravitational lensing 297–8
gravity 90–1, 133–4
    general relativity theory 82, 90, 142
    Newtonian 81–2, 90, 237
Great Comet 1811 279
Greek astronomy 24–5, 29–36, 37–8,
    60, 66, 82, 96, 144, 152, 276–7
greenhouse effect 18, 95–6, 112, 117

Hale-Bopp comet 280, 283, 283
Hall, Asaph 205, 211
Halley, Edmund 187, 282
Halley's comet 276–7, 278, 279–80,
    282, 291–2, 291
Harriot, Thomas 21
Hartley 2 (comet) 293, 294
Haumea 270, 271
Hayabusa mission 263
Heaven's Gate cult 280
Helene 222
heliocentric model 34, 38–45, 47,
    52–7, 87
helioseismology 74
helium 74, 79, 177, 201, 240
Herschel, Caroline 281
Herschel, John 244–5
Herschel, William 84, 99, 101,
    153, 179, 204–5, 234–7, 244,
    255, 281
Hevelius 98
Hipparchus 29–32, 33, 36
Hodgson, Richard 68–9
Holmes (comet) 285–6
Horrocks, Jeremiah 97–8
Hubble Space Telescope 233, 242,
    247, 261–2

Huygens, Christiaan 153, 179, 187,
203–4
Huygens lander 226–7
Hven observatory 46
Hyakutake (comet) 283
Hydra 269
hydrogen 74, 79, 177, 201, 240,
299–300
Hyperion 210, 225, 225

Iapetus 204, 210–11, 215, 223–4,
225
ice giants 19, 233–4
see also Neptune; Uranus
243 Ida 261, 262
India 149, 150
inferior planets 86–7, 101
see also Mercury; Venus
inner planets 17–19, 263–4
see also Earth; Mars; Mercury;
Venus
International Space Station 20–1, 75
Io 177, 184–9, 185, 188, 191–2, 196
volcanoes 191–2, 193
ion drives 267
iridium 260
Itokawa 263

Janus 225
Japan 263
JAXA 94
JUICE 200
Juno (asteroid) 256
Juno mission 199–200
Jupiter 17, 19, 57, 71, 156, 177–200,
271–2
and asteroids 257
atmosphere 177, 180–1, 190
belts and zones of 181–2, 181
clouds 179–80, 190, 198, 200
and comets 286, 288
differential rotation 179
features 180–3
Great Red Spot 19, 179, 180–1,
183, 190

impacts 178, 197–9, 199, 275, 303
missions to 189–94, 196,
198–200
moons 19, 57, 177–8, 184–9,
185, 188, 191–8, 193, 195, 200
observation 177–85, 195–8
opposition 184
orbit 16, 49, 50
ring system 196–7
storms 19, 177, 179–82, 190
water on 198
and the wobble of the Sun 297

Keck telescope 242, 245, 247
Kepler, Johannes 21, 41, 48–56,
50–1, 55, 97–8, 105, 107, 151,
156, 276
Kirchhoff, Gustav 78–80
Kohoutek's comet 282
Kreutz family (comets) 284–5
Kuiper, Gerard 206, 245, 248, 270
Kuiper Belt 20, 250, 251, 253, 254,
268–73, 269
comets 286, 288

Lascaux caves 24
Lassell, William 205, 236, 238, 244,
248
'Late Heavy Bombardment' 146–7,
217
lava 116–17, 147
Le Verrier, Urbain 237, 238, 252
Lexell, Johan 235, 237
libration 132
life
on Earth 120, 276, 288, 292,
299–304
extra-terrestrial 193–4, 220, 296,
300–2, 304
light, speed of 185–9, 188
'limb darkening' 205–6
Little Ice Age 77
Lowell, Percival 154–5, 268
lunar eclipses 142–3, 143, 186
Lutetia 263

Mab 247
Magellan 111, 113–14, 113
magma 94, 147
magnetic fields
  moons with 194–5
  planetary 117, 117–18, 244, 302
  of the Sun 60–1, 67–8, 71, 72,
    75
magnitude 32
Makemake 270, 271
Mariner probes 81, 93–6, 161–2,
  164, 172
Mars 17–18, 39, 94, 151–76, 162,
  169, 303
  and asteroids 257
  atmosphere 18, 153, 157, 163,
    174–6
  axis tilt 174
  canals of 154–5
  clouds 157, 159
  curse of 164–5
  dust storms 157, 161
  formation 121
  'four faces of' 158–60, 159
  journeys to 20–1
  life on 163–4, 167, 170–1,
    174–6
  maps of 154
  meteorites from 173, 261
  missions to 161–71
  observation 85, 86, 152–5,
    157–8
  opposition 156, 157
  orbit 16, 52, 53, 156
  as paradise lost 171–6
  plate tectonics 172–3
  polar ice caps 158, 173, 174
  rotational period 153
  seasons 157–8
  surface 94, 151–2, 158–62, 159,
    166–75
  temperatures on 18
  volcanoes 18, 160, 161, 172
  water on 18, 157, 161–3, 167–9,
    171–5
Mars Exploration Rovers 166–9
Mars Global Surveyor 164

Mars Odyssey 165
Mars Reconnaissance Orbiter 165,
  174–5
Mars Science Laboratory 169–71
mathematical astronomy 97
mathematics 29–30, 33, 39
253 Mathilde 262
'Maunder Minimum' 77
Maven spacecraft 174, 176
Mayans 96
McNaught (comet) 284
Mercury 17, 18–19, 64, 81–94, 97
  colour 89
  double sunrise of 92
  elongations 87, 88, 90
  formation 121
  iron core 93–4
  lack of an atmosphere 18–19, 81
  observation 81–9, 87, 91, 93–4
  orbit 16, 86, 87, 87, 90–3, 91
  phases of 84, 87–9, 87, 91
  rotation 92
  surface 94
  synodic periods 91
  temperatures on 19, 81
  volcanic activity 94
  water on 93
MESSENGER 81–2, 93–4
Messier list 281
Meteor Crater, Arizona 259, 259
meteorites 258–61, 259, 263–4
  Martian 173, 261
meteors 289–90
methane 175–6, 206, 226, 228, 240,
  250, 292
Methone 215
Milky Way 23
Mimas 204, 210, 214, 215, 218–19,
  219
'minor planets' 20, 253
Miranda 245, 246, 247
Moon 28–9, 34, 119, 122–4, 303
  craters 127–33, 146
  dark side of the 132–3
  eclipses 139–44, 141, 186
  far side of the 147–8
  formation 145–7

(Moon *continued*)

Full 123, 125, 126–7, 130–1, 142–3
gravity 133–4
missions to the 20, 148–50
motion 38
mountains/craters 100
New 29, 124, 125, 126–8, 139,
141–3
observation 127–33
orbit 92, 124–7, 125, 132–3,
138–40, 143–4
parallax 30–1
phases of the 124–8, 125
rotation 133, 138, 139
'seas' (*maria*) 122–4, 128–33, 131,
146–9
size 122
and solar eclipses 66
terminator 126–8
tides 111, 133–9, 136
water on 149–50
Moore, Patrick 30, 132–3

NASA 25, 72, 81, 111, 113–14,
149–50, 161–2, 164–6, 171, 173–4,
176, 199, 262–3, 265, 291, 293
NEAR-Shoemaker 262
Neptune 17, 19, 56, 156, 233–4,
271–2
and asteroids 257
atmosphere 240–4
discovery 237–8, 252, 255, 268
magnetic field 244
missions to 241–2
moons 238, 239, 248–52
observation 233–4, 239
orbit 16, 240, 244, 268
rings 248, 251–2, 251
seasons 240
storms 242, 243
structure 240–4
Nereid 248
New Horizons mission 21, 193, 196,
272–3
Newton, Isaac 54, 81–2, 97, 99, 187,
237, 278

Nix 269
nuclear fusion 15, 74

Oberon 236, 239, 245
observation 28–33, 39, 49–50
occultation 30, 212–13
Oort cloud 286–7, 288
Opportunity (rover) 166–9
opposition 156, 157, 184, 209–10,
239
optical filters 89, 104, 184, 210
solar 64, 65
orbits
circular 34–5
elliptical 34–5, 52–6, 55, 57, 87,
90, 132, 140, 144, 151, 156
synchronous 86
'organic chemicals' 300, 302
outer planets 17, 19–20, 156
see also Jupiter; Neptune; Saturn;
Uranus
oxygen 299–300

Pallas 256
Pallene 215, 216
Pan 214, 225
Pandora 214, 216, 218, 225
parallactic instruments 46
parallax 30–1
Pathfinder mission 164, 166
penumbra 140–1, 141, 143
perihelic precession 90–1, 91
periods 54
Phobos 158
Phoebe 215, 216, 223–4,
226
Phoenix mission 164
photosynthesis 299
Piazzi, Giuseppe 255, 256
Picard, Jean 186–7
pinhole cameras (camera obscuras)
60, 62, 63
Pioneer missions 177, 189, 213
planetary motion 34–5, 36, 49–50,
107

retrograde 41, 52–4, 53, 109–12, 248, 249
planets 27, 34
  formation 110, 121, 255, 263–4, 271
  habitable 301–2
  search for 296–9, 303
plasma (solar) 73–4
Plough (Great Bear) 24–5, 26
'plurality of worlds' 43
'Plutinos' 272
Pluto 21, 196, 253, 268–73
Polaris 31
pole star 31
Polydeuces 222
Pons, Jean-Louis 281
potassium 94
Prometheus 214, 216, 225
Proteus 249
Ptolemy, Claudius 32, 34–9, 36, 41, 44, 57, 97, 185, 202
Pythagoras 34

1992 QB1 270
quadrants 46

radar 113–14
radiation 75–6, 118
rain, sulphuric acid 112
religion 27, 34, 36
Renaissance 37
Rhea 204, 210, 220 3, 226
Rheticus, Georg Joachim 40
Rømer, Ole 185–9, 188
Rosetta spacecraft 263, 293–4
Russia 176

Saros cycle 144
satellites, solar observation 71–2, 75
Saturn 17, 19, 20, 39, 156, 177, 189, 196, 201–32
  atmosphere 201, 208
  and comets 286, 288
  features 207–10, 208

missions to 201, 213–15, 218–20, 222–4, 226–32, 227
moons 201, 204–6, 210–11, 214–30, 219, 221, 241
observation 201–12
opposition 209–10
orbit 16, 49, 50
ring system 19, 201, 203–5, 207–10, 208, 212–18, 214, 224–5, 230
rotational period 205
seasonal aspects 206, 206, 207, 230, 232
storms 201, 202, 206, 211–12, 230–2, 231
Schiaparelli, Giovanni 85–6, 89, 154
Schröter, Johann Hieronymus 84–5, 99–101, 205
Schröter effect 100, 102, 104
science 37
seasons 120, 157–8, 206–7, 206, 230, 232, 240–3, 241
Sedna 273–4
seeing 89, 103, 156, 184, 209
sextants 46
shepherd moons 214, 218, 251
Shoemaker Levy 9 (comet) 197–8, 199, 275
sidereal years 31
SOHO satellite 71
Solá, Josep Comas 205–6
solar cycles 61, 68, 76
Solar Dynamics Observatory 72
solar eclipses 64, 66, 140, 141, 144
solar flares 68–9, 75
solar storms (coronal mass ejections) 68–9, 72, 75–6
Solar System
  defining the 15–22
  geocentric model 34–6, 36, 37
  heliocentric model 34, 38–45, 47, 52–7, 87
  planetary orbits 16
  size of the 107–9
solar tides 138
solar wind 68, 71

sound waves 73–4
space weather 75–7
spectroscopy 59, 78–80, **78**, 279
Spirit (rover) 166–7, 169
Stardust probe 292, 293
stars 295–8
  binary 298–9
  composition 79–80
  double 235
  magnitude 32
  wobble 297, 298
  see also specific stars
STEREO mission 72
Sun 15–17, **16**, 29, 30, 39, 59–80,
  295
  and the aurorae 69–70
  central temperature 15
  as the centre of the Solar System
    34, 38–45, 47, 52–7, 87
  and comets 284–5, **285**
  composition 79–80
  convection cells 74
  corona 66–8, 72, 140
  diameter 73
  and the Earth 121, 139
  energy of the 74, 76
  and the equinoxes 31
  from space 70–2
  interior 73–4
  and life 303
  magnetic field 60–1, 67–8, 71,
    72, 75
  and Mars 156
  mass 68
  and Mercury 90, **91**, 92
  observation 62–5, 70–2, 78–80,
    89
  as orbiting the Earth 34
  photosphere 66–7, 72
  and planetary motion 52–4, 53,
    55
  prominences 65
  size 15, 66, 73
  and space weather 75–7
  spectroscopy 59, 78–9, **78**
  sunspots 60–2, **61**, 64, 68, 76–7
  surface temperature 15

weight 15, 68, 74
wobble 297
see also solar...
sunlight 74
supernovas 46–7
superstition 27, 37, 45, 276–7,
  279–80

Taurus 38, 235
tectonic activity 115, 120–2, 172–3,
  218–19, 250
telescopes 21–2, 56–8, 60, **98**
  and Jupiter 177–85, 195–8
  and Mars 153–5, 157–8
  and Mercury 83–6, 87, 89, 91
  and the Moon 128–30, 132
  and Neptune 233–4, 239
  reflecting 63, 99, 100, 179, 235
  refractors 83, **84**, 97, 179
  and Saturn 201–12
  solar 65
  and the Sun 62–4, 70–1
  and Uranus 233–4, 239, 242–3
  and Venus 97–109
Telesto 222
Tempel I (comet) 292–3
Tethys 204, 220–2, 310
Thermisto 196
tholins 224, 250
thorium 94
tides 111, 133–9, **136**, 192, 221–2
Titan 201, 204–6, 210, 218, 226–9,
  232, 241
  atmosphere 226
  cryovolcanism 229
  hydrocarbon lakes 226–8, **227**
  interior 228–9
  life on 229
  orbit 228
  temperatures on 226
Titania 236, 239, 245
Titius-Bode Law 254–5
Tombaugh, Clyde 268
Toutatis 263
transit 97–8, 102, 105–9, **106**, 184,
  296–8

trigonometry 29, 31
Triton 238, 239, 248, 249–51
  cryovolcanism 250–1, 272
Trojan asteroids 257
Trojan moons 222

Ulysses (satellite) 71
umbra 140–1, 141, 143
Umbriel 245
Universe
  centre 34–5, 36, 37
  infinity 43
  theories of the 33–6, 37–47, 52–8
Uranus 17, 19, 56, 84, 156, 233, 271
  and asteroids 257
  atmosphere 240–4
  discovery 235–7, 255
  infrared images 242, 245
  magnetic field 244
  missions to 241–2
  moons 236, 239, 244–7, 245, 247
  observation 233–4, 239, 242–3
  orbit 16, 240–1, 241, 244, 268
  rings 236–7, 246–8, 249
  seasons 240–3, 241
  storms 242
  structure 240–4

Vega spacecraft 291
Venera probes 113
Venus 17, 18, 57, 64, 87, 93, 95–118, 152, 189, 303
  Ashen Light 104–5
  atmosphere 18, 95–6, 109, 111–13, 117–18
  common features 103
  craters 114
  formation 121
  Himalayas of 101
  ionosphere 118
  lack of water 116, 117
  magnetic field 117–18
  observation 88, 96–109, 112–14
  orbit 16, 101–2, 105–7, 106

  phases 84, 101–2
  rotation 109–12
  sulphur dioxide clouds 109, 112, 116
  surface 95, 113–17
  synodic cycle 102
  temperatures on 18, 112
  tides 111
  transits 102, 105–9, 106
  volcanoes 112–17
Venus Express 111, 116
Vesta 254, 256, 263–5, 264
Viking probes 160, 162–4, 166, 169, 172, 175
volcanoes 18, 94, 112–17, 121–2, 160–1, 172, 191–2, 193, 229, 250–1, 272
  see also cryovolcanism
Voyager missions 177, 180, 189–92, 196, 213–14, 218, 222–3, 226, 233, 238, 241–2, 243, 244–6, 249–51
VT-2004 108

water 149–50
  on Earth 116, 120, 276, 288, 299–300
  on other planets 18, 93, 157, 161–3, 167–9, 171–5, 198
West (comet) 282
Wild 2 (comet) 292
WISE telescope 265, 274
Wollaston, William 78
Wratten number 89

Picture Credits

BBC Books would like to thank the following individuals and organizations for providing photographs and for permission to reproduce copyright material. While every effort has been made to trace and acknowledge copyright holders, we would like to apologize should there be any errors or omissions.

BBC p12; Dr Paul Abel p26; New York Public Library/Science Photo Library p28; SOHO/MDI (ESA & NASA) p61; Emilio Segre Visual Archives/American Institute of Physics/Science Photo Library p78; Asadollah Ghamari p106; NASA/JPL p113, p243, p251; NASA p123; Pete Lawrence p131; Viking Project/USGS/NASA p162; NASA/JPL-Caltech p169; NASA/JPL-Caltech/MSSS p171; Nick Damico p185; NASA/Johns Hopkins University Applied Physics Laboratory/Southwest Research Institute p193; Galileo Project/JPL/NASA p195; Hubble Space Telescope Comet Team/NASA p199; NASA and The Hubble Heritage Team (STScl/AURA) Acknowledgement: R.G. French (Wellesley College), J. Cuzzi (NASA/Ames), L. Dones (SwRI) and J. Lissauer (NASA/Ames) p206; CICLOPS/JPL/ESA/NASA p214; NASA/JPL/Space Science Institute p219, p221, p225; NASA/JPL-Caltech/ASI p227; NASA/JPL-Caltech/Space Science Institute p231; Lawrence Sromovsky, Pat Fry, Heidi Hammel, Imke de Pater/University of Wisconsin-Madison p245 (top); NASA/JPL/STScl p245 (bottom); NASA/JPL/USGS p247; NASA/USGS p259; NASA/Jet Propulsion Laboratory p262; NASA/JPL-Caltech/UCAL/MPS/DLR/IDA p264; Andy Gannon p283; ESA p291; NASA/JPL-Caltech/UMD p294

The following illustrations by Richard Palmer Graphics were based on originals drawn by Dr Paul Abel: p103, p159, p181